D0876228

THE DYNAMICS OF DISABILITY

MEASURING AND MONITORING DISABILITY FOR SOCIAL SECURITY PROGRAMS

Gooloo S. Wunderlich, Dorothy P. Rice, and Nicole L. Amado, *Editors*

Committee to Review the Social Security Administration's
Disability Decision Process Research

Board on Health Care Services
INSTITUTE OF MEDICINE

and

Committee on National Statistics
Division of Behavioral and Social Sciences and Education
NATIONAL RESEARCH COUNCIL

NATIONAL ACADEMY PRESS
Washington, D.C.

NATIONAL ACADEMY PRESS • 2101 Constitution Avenue, N.W. • Washington, DC 20418

NOTICE: The project that is the subject of this report was approved by the Governing Board of the National Research Council, whose members are drawn from the councils of the National Academy of Sciences, the National Academy of Engineering, and the Institute of Medicine. The members of the committee responsible for the report were chosen for their special competences and with regard for appropriate balance.

Support for this project was provided by Contract No. 600-96-27893 between the National Academy of Sciences and the Social Security Administration. Support of the work of the Committee on National Statistics is provided by a consortium of federal agencies through a grant between the National Academy of Sciences and the National Science Foundation (Grant No. SBR-9709489). The views presented in this report are those of the Committee to Review the Social Security Administration's Disability Decision Process Research and are not necessarily those of the funding agency.

Additional copies of this report are available for sale from the National Academy Press, 2101 Constitution Avenue, N.W., Box 285, Washington, DC 20055. Call (800) 624-6242 or (202) 334-3313 (in the Washington metropolitan area), or visit the NAP's home page at **www.nap.edu**. The full text of this report is available at **www.nap.edu**.

For more information about the Institute of Medicine, visit the IOM home page at: **www.iom.edu**. For more information about the Committee on National Statistics, visit the CNSTAT home page at **www2.nas.edu/cnstat**.

THE NATIONAL ACADEMIES

National Academy of Sciences
National Academy of Engineering
Institute of Medicine
National Research Council

The **National Academy of Sciences** is a private, nonprofit, self-perpetuating society of distinguished scholars engaged in scientific and engineering research, dedicated to the furtherance of science and technology and to their use for the general welfare. Upon the authority of the charter granted to it by the Congress in 1863, the Academy has a mandate that requires it to advise the federal government on scientific and technical matters. Dr. Bruce M. Alberts is president of the National Academy of Sciences.

The **National Academy of Engineering** was established in 1964, under the charter of the National Academy of Sciences, as a parallel organization of outstanding engineers. It is autonomous in its administration and in the selection of its members, sharing with the National Academy of Sciences the responsibility for advising the federal government. The National Academy of Engineering also sponsors engineering programs aimed at meeting national needs, encourages education and research, and recognizes the superior achievements of engineers. Dr. Wm. A. Wulf is president of the National Academy of Engineering.

The **Institute of Medicine** was established in 1970 by the National Academy of Sciences to secure the services of eminent members of appropriate professions in the examination of policy matters pertaining to the health of the public. The Institute acts under the responsibility given to the National Academy of Sciences by its congressional charter to be an adviser to the federal government and, upon its own initiative, to identify issues of medical care, research, and education. Dr. Harvey V. Fineberg is president of the Institute of Medicine.

The **National Research Council** was organized by the National Academy of Sciences in 1916 to associate the broad community of science and technology with the Academy's purposes of furthering knowledge and advising the federal government. Functioning in accordance with general policies determined by the Academy, the Council has become the principal operating agency of both the National Academy of Sciences and the National Academy of Engineering in providing services to the government, the public, and the scientific and engineering communities. The Council is administered jointly by both Academies and the Institute of Medicine. Dr. Bruce M. Alberts and Dr. Wm. A. Wulf are chairman and vice chairman, respectively, of the National Research Council.

EDWARD H. YELIN, Professor of Medicine and Health Policy, Department of Medicine and Institute for Health Policy Studies, University of California at San Francisco

Study Staff

GOOLOO S. WUNDERLICH, Study Director
NICOLE AMADO, Research Associate

DONALD STEINWACHS, Professor and Chair, Department of Health Policy and Management, Bloomberg School of Public Health, Johns Hopkins University, Baltimore, Maryland

PAUL C. TANG, Medical Director, Clinical Informatics, Palo Alto Medical Foundation, Palo Alto, California

JANET M. CORRIGAN, Director, Board on Health Care Services, IOM

Acknowledgments

The Committee to Review the Social Security's Disability Decision Process Research gratefully acknowledges the contributions of the many individuals and organizations over the course of six years, not all of whom can be individually listed here, who participated and gave generously of their time and knowledge to this study.

Support for this study was provided by the Social Security Administration (SSA). Staff of SSA was helpful in providing information about the research projects undertaken relating to the disability decision process and the planning, development, and statistical design of the National Study of Health and Activity, the two principal study areas of the contract. We particularly wish to thank Scott Muller, Senior Economist, Office of Research Evaluation, and Statistics, who served as the SSA project officer throughout the duration of the study. We also acknowledge David Barnes and Rosanne Hanratty who served as co-project officers with Dr. Muller in the early years of the study. In addition we acknowledge the many federal and nonfederal government officials and those from the research and disability communities who gave expert presentations to the committee at its meetings and those who participated in the two large workshops organized by the committee. They are listed in Appendix A and B of this report.

We are grateful to Elizabeth Badley, Alan Jette, Cille Kennedy, Nancy Mathiowetz, Laura Trupin, and Edward Yelin, the authors of the commissioned papers prepared for the study. These papers were used extensively by staff and committee in drafting the report. These papers are included in Part II of the report.

We acknowledge with gratitude the contributions of our consultant, Nancy Mathiowetz, Associate Professor at the University of Maryland's Joint Program on Survey Methodology. Dr. Mathiowetz provided assistance and technical expertise to staff and the committee in organizing the workshop on Survey Measurement of Work Disability, co-editing the summary report of the workshop, and advising on dissemination of the report. She also authored two of the commissioned papers on statistical issues associated with survey measurement of disability and the various possible methods of obtaining continuing information on disability issues confronting SSA.

We acknowledge the contributions of the IOM study staff to whom an important debt of gratitude is owed. The committee is especially grateful for the guidance and efforts of our study director, Gooloo S. Wunderlich, for the enormous contribution she made to the study. She had primary responsibility for organizing the deliberations of the committee and preparing the drafts of the study reports, tasks that she accomplished with considerable skill and tact. Gooloo's professionalism, knowledge, and extraordinary commitment and perseverance were critical to resolving many policy and technical issues and completing the study. Nicole Amado served ably as a senior project assistant, research assistant, and more recently as research associate—at times simultaneously handling the functions of all three positions. Her excellent support and attention to detail were critical to the success of the final report. She independently developed all of the components required in IOM reports and reviewed and edited the various drafts of the report for accuracy and style under very tight deadlines. She researched and developed all the tables and figures in the report and reformatted all the tables in Part II of the report. The committee acknowledges the several other staff who assisted the committee at various times and in varying capacities over the six years of the study.

Other IOM staff provided support and assistance to the study committee. The IOM Financial Associates helped keep our budget in order. The Staff of the Office of Reports and Communication handled the logistics of the reports review process and shepherded the reports through the editing and production process. Sally Stanfield at the National Academy Press was supportive and helpful as always for each of the six reports issued by the committee.

Finally, I would like to thank all the members of the committee for their generous contribution of time and expert knowledge to the deliberations and the preparation of the committee reports including this final report.

Dorothy P. Rice, *Chair*
Committee to Review the Social Security
Disability Decision Process Research

Reviewers

The report has been reviewed in draft form by individuals chosen for their diverse perspectives and technical expertise, in accordance with procedures approved by the National Research Council's Report Review Committee. The purpose of this independent review is to provide candid and critical comments that will assist the institution in making its published report as sound as possible and to ensure that the report meets institutional standards for objectivity, evidence, and responsiveness to the study charge. The review comments and the draft manuscript remain confidential to protect the integrity of the deliberative process. We wish to thank the following individuals for their review of this report:

Henry Aaron, Senior Fellow, Brookings Institution, Washington, D.C.
Richard G. Frank, Professor, Harvard Medical School, Boston, Massachusetts
David Gray, Researcher Professor, Washington University School of Medicine, St. Louis, Missouri
Robert Haveman, Professor, University of Wisconsin—Madison
Lisa Iezzoni, Professor of Medicine, Harvard Medical School, Boston, Massachusetts
Corinne Kirchner, Director, Policy Research and Program Evaluation, American Foundation for the Blind, New York, New York
James Morgan, Professor and Research Scientist Emeritus, University of Michigan
Janet L. Norwood, Consultant, Chevy Chase, Maryland

Although the reviewers listed above have provided many constructive comments and suggestions, they were not asked to endorse the conclusions or recommendations nor did they see the final draft of the report before its release. The review of this report was overseen by Edward B. Perrin, Professor Emeritus, University of Washington and Senior Scientist, VA Puget Sound Health Care System, appointed by the Institute of Medicine, and Joseph P. Newhouse, Professor, Harvard Medical School, appointed by the National Research Council's Report Review Committee. They were responsible for making certain that an independent examination of this report was carried out in accordance with institutional procedures and that all review comments were carefully considered. Responsibility for the final content of this report rests entirely with the authoring committee and the institution.

Contents

Tables and Figures

TABLES

FIGURES

Summary

Disability is a dynamic process that can fluctuate in breadth and severity across the life course and may or may not limit ability to work. Disability is not a static event because it is the adaptation of a medical condition in the environment in which one lives. It needs to be monitored, measured, and evaluated on a regular basis to understand the growth in the Social Security Administration's (SSA) disability programs, estimate the current and future prevalence of disability, ensure an effective and efficient system of determining program eligibility, and maintain fiscally responsible administration of the programs.

During the past two decades, the Social Security Disability Insurance (SSDI) program and the Supplemental Security Income (SSI) program have experienced an unexpected, rapid growth.[1] More people are receiving disability benefits today than ever before. With the exception of a period in the late 1970s and in the early 1980s, the number of beneficiaries on the rolls has increased steadily as the growth in awards has outpaced terminations. The mix of beneficiaries also has been changing. In the past, people entering the programs were more likely to be over 50 years of age

---

[1]SSDI is an insurance program that provides payments to persons with disabilities based on their having been covered previously under the Social Security program. SSI is a means-tested income assistance program for disabled, blind, and aged persons who have limited income and resources regardless of their prior participation in the labor force. The definition of disability and the process of determining disability are the same for both programs.

suffering from conditions of the circulatory system or disabling musculo-skeletal conditions. In recent years new beneficiaries are more likely to be younger and have mental impairments. They are likely to remain on the rolls longer.

Many factors have shaped the disability programs over the years, including economic conditions; the changing nature of work; the maxi-mum level of gainful activity allowed for people on disability; the incen-tives and increased outreach by SSA and disability advocates; legislative actions, court decisions, and administrative initiatives undertaken by SSA in the way disability decisions are made; public perception about the ease of qualifying for benefits; eligibility for medical benefits through Medi-care or Medicaid; demographic composition and characteristics of the population; and the types of impairments of applicants that are recog-nized for disability cash benefits. However, the impact of any one factor on the demand for, and the provision of, disability benefits is difficult to determine definitively in the absence of data.

The challenge for SSA is to acquire these data and use them to man-age the disability programs more effectively. The disability rolls are pro-jected to grow over the coming decades as the baby boom generation reaches the age of increased likelihood of developing disabilities. The gradual increase from 65 to 67 in full retirement age also means that disabled workers may remain on the disability rolls for two additional years before converting to the old age survivor benefits. Ongoing and future research using new data sources should provide information about current enrollment in disability programs and allow projections of future growth and program costs. The results should lead to clearer and more workable policies, rules and guidelines to operate its programs.

STUDY SCOPE

In 1996, SSA requested that the Institute of Medicine (IOM), in col-laboration with the National Research Council's Committee on National Statistics (CNSTAT) conduct an independent review of the statistical de-sign and content of the disability survey under development (the Na-tional Study of Health and Activity (NSHA)) and of its research plan for the redesign of the disability decision process. The committee's specific tasks include, but are not limited to, the following:

- review the scope of work for the NSHA request for proposal and the design and content of the survey as proposed by the survey contractor;
- review and evaluate the preliminary design of the NSHA (the pro-tocol developed by Westat, Inc.) and subsequent modifications

made by SSA. Identify statistical design, methodological, and content concerns, and other outstanding issues, and make recommendations as appropriate;

- review SSA's research plan and time line for developing a new decision process for disability, and offer comments and recommendations on direction to the research; and
- review all completed research including, but not limited to, reviewing research into existing functional assessment instruments conducted under contract to SSA by Virginia Commonwealth University, and provide advice and recommendations for adopting or developing functional assessment instruments or protocols for the redesigned disability process and NSHA.

Realizing that some key components of the research and testing of the NSHA design were behind schedule, SSA extended the original four-year contract period for an additional two years. Also, SSA informed the committee in late 1999 that it would no longer pursue the development of the new decision process as proposed in its disability redesign plan; instead it had decided to focus on improving the current process.

To meet its responsibilities, the committee reviewed an extensive body of research literature and other relevant reports, heard from a number of experts in the field, and commissioned several background papers from experts. The committee held two large workshops to obtain input from a wide range of researchers and other interested members of the public and to augment its knowledge and expertise by more focused discussion on issues of functional capacity and work requirements and survey measurement of work disability. Throughout the study the scope and extent of the committee's review has depended on what was initiated or completed and what SSA made available to the committee.

The committee's review of the NSHA and the research plan for the redesign of the disability decision process represent two separate areas of study. For the most part, therefore, they are discussed separately in this report. Also, the review of the redesign initiative is limited to reviewing the research under way and planned for the redesign of the disability *decision* process, which is only one element of SSA's total effort to reengineer the disability *claims* process.

Recognizing the need to provide SSA with timely advice, the committee issued three interim reports with detailed technical recommendations on topics that needed immediate attention by SSA (see Chapter 1).

The committee's major findings and conclusions based on this review and its deliberations are summarized below, followed by the text of all the recommendations.

FINDINGS AND CONCLUSIONS

Conceptual Issues in Defining Work Disability

There is no agreement on the definition and measurement of disability. The meaning assigned to the term depends on the uses to be made of the concepts. SSA's focus in both the SSDI and the SSI programs is on work-related disability, as defined in the Social Security Act. It defines disability (for adults) as the inability to engage in any substantial gainful activity by reason of any medically determinable physical or mental impairment that can be expected to result in death or that has lasted, or is expected to last, for a continuous period of not less than 12 months. An individual's physical and mental impairment(s) must be of such severity that he or she not only is unable to do the previous work but cannot, given the person's age, education, and work experience, engage in any other kind of substantial gainful work that exists in the national economy, regardless of whether such work exists in the local area, or whether a specific job vacancy exists, or whether the person would be hired if he or she applied for work.

The Social Security definition of disability was developed in the mid-1950s at a time when a greater proportion of jobs was in manufacturing and more required physical labor than today. It was therefore expected that people with severe impairments would not be able to engage in substantial gainful activity. Over the years, many changes have occurred: the nature of work has shifted from manufacturing toward service industries; medical and technological advances have made it possible for more severely disabled persons to be employed; and in recent years, public attitude also has changed as reflected in the enactment of the Americans with Disabilities Act of 1990 (ADA).

In recent years the concept of disability has shifted from a focus on diseases, conditions, and impairments per se to more on functional limitations caused by these factors. Critics suggest that the SSA's definition of disability and its process for determining program eligibility have not kept pace with the changes. The committee recognizes the administrative difficulties involved in paying more attention in the disability determination process to the physical and social factors in the work environment. Moreover, it might require major shifts in the orientation of the Social Security disability programs to ways to influence the environment in which the applicant might work and to "return-to-work" activities, and might ultimately involve changes in SSA's implementing regulations.

Work Disability Monitoring System

Managing SSA's disability programs and adapting them to the evolving needs of Americans with disabilities require ongoing data collection with instrumentation that can be updated to reflect any future changes in either conceptual and measurement issues or SSA's eligibility protocol. Only then can analysts and policymakers have the information necessary to understand and predict the impact of changes in the environment on an individual's propensity to apply for benefits. *The committee therefore recommends that the Social Security Administration develop an ongoing disability monitoring system, building from its experience with the NSHA.* Such a monitoring system should consist of (1) a periodic, comprehensive, and in-depth survey to measure work disability; and (2) a small set of core measures in the intervening years derived from other surveys, reinterviews, and/or administrative data.

Such a system should provide SSA with data needed to respond to a variety of policy and planning issues including, but not limited to, the following: the size, distribution, and characteristics of the working population with disabilities; demographic trends; labor market dynamics; changes in economic conditions; needs of minority and special populations with disabilities; quality of life for disabled workers; functional status of people with disabilities; role of the states; and legislative, regulatory, and judicial impacts on disability programs.

To ensure the utility of a monitoring system for policy decisions and implementation SSA should establish a clear set of information objectives in developing the substantive content of the monitoring system. The design of a disability monitoring system must consider the information needs of the system and the impact of alternative design options on meeting analytic goals. These options can be arrayed along lines of richness of the data, quality of the data, and costs. Resources and changing policy needs may dictate many of the system's design features and selection of the design options and might include

- sponsoring surveys at frequent intervals based on self-report data from a reduced set of disability measures;
- funding additional survey questions, suitable for estimation of the size of the population eligible for disability benefits, as part of, or supplement to, an ongoing household survey;
- conducting longitudinal data collection;
- forming a partnership with other ongoing surveys;
- linking survey information with administrative databases; and
- conducting ad hoc special studies on specific emerging policy issues and to explore other questions that do not need continuing

data. For example, a follow-up survey of applicants and a survey of employers and people with disabilities might be conducted to obtain information on accommodations that employers provide.

To develop a monitoring system SSA needs to set up a multiyear planning period to systematically design and establish the monitoring system. The monitoring system should be designed with sufficient flexibility to accommodate the evolving medical, legal, social, and policy needs; to make the best use of the design and data from existing federal surveys; and to ensure the availability of sufficient qualified research staff to design and oversee the system.

The committee recognizes that despite its many benefits, developing and implementing the recommended disability monitoring system raises several important issues that would require careful examination and resolution. Many of these issues relate to conceptual definition, method, timing, collaboration with partner agencies, and resource requirements. *The committee suggests a phased three-year planning period starting in 2003. Also, SSA should establish a continuing, external group of technical experts for the planning and implementation of the recommended disability monitoring system.*

Survey Measurement of Disability

The National Study of Health and Activity is a complex, national, sample survey designed to estimate the number and characteristics of a broad range of working age people with disabilities that affect their ability to work and carry out activities of daily living. SSA has contracted with Westat to conduct the survey. In its review of NSHA, the committee focused on measurement issues and on the adequacy of the research design and its implementation plan.

Time Demands to Achieve Survey Quality

Careful survey design and measurement require considerable development and field-testing prior to implementation. Cost savings that appear to arise when work is rushed are illusory. The original schedule for the conduct of NSHA did not permit deliberate and rigorous decisions about revisions of the design, procedures, or questionnaire content. The rush to launch the national survey has caused serious logistical inflexibility during the various phases of the survey.

An example of inadequate time for the developmental work and testing is the pilot study. SSA originally planned to complete developmental work and conduct the pilot six to nine months after award of the contract for the survey. In response to a recommendation by the committee in its first interim report, SSA decided to conduct a large, comprehensive pilot study. However, little time was allowed in the schedule for research, development, testing, and making modifications. Decisions had to be made throughout the process, and it became obvious that insufficient time was allowed to resolve issues and test alternatives. As a result of the pilot study experience, the data collection plans are being restructured, and the mode of data collection changed because of poor results with the random digit dialing (RDD) sampling frame.

The committee understands that SSA is already addressing many of the issues raised by the committee in this report. It notes that recently SSA has approved significant additional time to the schedule to evaluate the results of the pilot study and to test alternative solutions to problems before starting the national study.

Not allowing sufficient time for research, development, and testing prior to launching a major complex survey has resulted in the need to repeatedly revise the timetable for developing and conducting the survey. The most recent revised schedule available to the committee called for the "end-to-end" test data collection from December 2001 to February 2002; dress rehearsal data collection from December 2002 to January 2003; and the main study to start early in 2003. Thus, the survey originally planned for mid-2000 is now scheduled to start in 2003 and assumes a multiyear data collection plan.

Issues Associated with Survey Measurement of Disability

The experience to date with NSHA and other similar surveys indicates that measurement issues related to work disabilities are complex. The complexity stems, in part, from differences in conceptual models of the enablement–disablement process and alternative interpretations of the various conceptual models. The various constructs do not necessarily identify the same population.

The committee underscores the need for the development of objective measures of both the physical and the social environment. Toward this end, the committee notes the need to develop and test questions concerning social climate, barriers, and stigma. While these questions are especially important for those with mental illness, they are relevant also for, and should be asked of, all persons with disabilities.

Larger samples reduce the uncertainty that the survey results will depart from those in the full target population. Since the committee's first interim report, it has raised questions about the adequacy of the sample size targets and especially about the allocation of people among the four subgroups established by SSA—nonbeneficiaries with severe disabilities, persons with significant but lesser impairments, nondisabled persons, and current beneficiaries. The committee is concerned that the targeted sample sizes may not support SSA's requirements for estimation and analytical purposes. It has not seen the logic behind these targeted sample sizes. The rationale and plans for analysis were never provided to the committee.

In its plans for achieving the targeted sample sizes, SSA has assumed response rates of about 90 percent for the various components of the NSHA. The committee believes that the expected response rate may be overly optimistic, especially for a population with disabilities. Even if these planned sample sizes can be achieved, the cells very likely will be much too small, especially if SSA stratifies for analytical purposes on more than one disabling condition and/or demographic and socioeconomic characteristics such as age, gender, and minority status, or working non-beneficiaries with specific disabling conditions. The committee has recently learned that SSA is rethinking these targets based on the evaluation of the pilot study results.

The committee has repeatedly stated in its interim reports and again in this final report that the NSHA, if well designed, could be the cornerstone for long-term disability research. However, the value of the information from any cross-sectional survey diminishes with time. *It is, therefore, critical that SSA update the comprehensive database regularly.*

Improving the Disability Decision Process

The goal of SSA's research plan to redesign the disability decision process was to devise a more efficient and more accurate method for making timely determinations of disability for Social Security claimants. Early in the study, the committee conducted a preliminary review of the general features and directions specified by SSA in its research plan and of the individual research projects completed and under way within the research plan. In an interim report of its findings, conclusions, and recommendations, the committee recommended that SSA adopt a rigorous research design process and develop, early in the research, measurable criteria and validation plans to enable SSA to make the ultimate judgments on whether or not the proposed changes would yield the desired results.

SSA concurred with some of the committee's conclusions and recommendations. However, rather than undertaking the measures recommended by the committee, SSA decided it would no longer actively pursue the redesign of the disability decision process. Now it plans to improve the current process, focusing at this time on updating the medical listings.

Unfortunately, the committee finds that several of the key issues it had identified with regard to SSA's research plan for redesigning the disability decision process must also be addressed in activities undertaken to improve the current process. Therefore these issues are reemphasized in this report.

Baseline Evaluative Criteria

Regardless of whether SSA attempts to redesign and develop a new disability decision process or leaves the current process in place and makes changes within the individual components of the sequential process, it needs to establish measurable criteria for assessing the current process. Data should be analyzed in the context of the established criteria in order to identify the nature of the problems in the current process. Although there is no "gold standard" for identification of individuals who are eligible for disability benefits, the committee recognizes that some criteria are needed to assess how accurate are the current determinations of disability. In reviewing the research proposals and other documents related to the redesign the committee found no indication that SSA had conducted any baseline analysis with predetermined criteria. Unfortunately SSA appears to be going down the same path now. *The committee reiterates its earlier position and recommends that before making the changes in the current decision process, SSA should establish evaluative criteria for measuring the performance of the decision process, conduct research studies and analyses to determine how the current processes work relative to these preestablished criteria, and then evaluate the extent to which change would lead to improvement.*

Since SSA is devoting its attention to updating the Medical Listings component of the decision process, this recommendation is most applicable to the Listings. However, the committee notes that the Medical Listings apply only to one step (step 3) of the five-step sequential evaluation process for determining disability. *The baseline evaluation should ultimately evaluate the total process and not just one component.*

Assessing Vocational Capacity

The Dictionary of Occupational Terminology (DOT) has served as a primary tool for determining whether a claimant has the capacity to work. However, the Department of Labor (DOL) is no longer updating the DOT. Although the replacement classification system, the Occupational Information Network (O*NET), will be very useful for DOL's purposes, it will not meet SSA's needs to define the functional capacity to work without major reconstruction. *Barring some resolution, SSA will be left with no objective basis upon which to justify decisions concerning an individual's capacity to do jobs in the national economy.* SSA might be cast back into the era in which it relied extensively on the testimony of "vocational experts" or their written evaluations.

Moreover, SSA has not updated the research base on the effect of *age, education, and work experience* on work disability. The research base was used in developing the medical–vocational guidelines of 1978. Since then much has changed with regard to the relative importance of each of these factors. As part of the initiative to redesign the decision process, SSA included in its redesign research plan an evaluation of the effect of vocational factors—age, education, and work experience—on the ability to work or adapt to work in the presence of functional impairment. A review of existing knowledge concerning vocational factors and their impact on the ability to perform jobs in the national economy raised challenging questions about the continuing validity of the approach taken by SSA's existing regulations. The review suggested a critical need for research designed to validate the use of vocational factors in SSA's disability decision process.

Disability Allowances

Over the past two decades, the number of disability beneficiaries in the working age population has risen steadily. Although the number of applicants for benefits has increased only moderately, the number of new beneficiaries has nearly doubled. Disability allowance rates (awards as a percentage of applications) have varied over time from 31.4 percent in 1980 to nearly 47 percent in 2000.

Variations in allowance rates occur for several reasons. For example, SSA's standards for judging claims differ over time. Dramatic reductions in allowance rates occurred when standards were abruptly tightened in 1980 and then subsequently relaxed. Significant differences are observed in allowance rates across states, between Disability Determination Service (DDS) decision makers, and between DDSs and administrative law judges. The allowance rate is also influenced by legislative changes as well as court decisions, and the adequacy of resources to process and review cases. *Increased research is needed to explain these variations and whether they are predictable.*

The objectives of the current disability decision process are to attempt to make decisions that are consistent with the statutory definition of disability as consistently, expeditiously, and cost-effectively as possible. Recent legislation—the Ticket to Work and Work Incentives Improvement Act of 1999 (P.L. 106-170)—suggests that Congress is increasingly interested in the return to work model and is prepared to have SSA experiment with some alternative strategies that might facilitate the pursuit of work rather than benefits. *The committee concludes that SSA should ini-*

tiate a research program for testing decision process models that emphasizes rehabilitation and return to work.

Enhancing SSA's Research Capacity

Throughout this report the committee recommends major research efforts to understand disability programs' growth and to effectively and efficiently administer these programs. Such research includes research on the measurement of work disability in a survey context, evaluation of the role of the environment and vocational factors in determining work disability, evaluation of functional capacity of applicants for disability benefits, and testing decision process models that emphasize rehabilitation and return to work. *The committee emphasizes that without the infusion of new resources, in terms both of dollars and recruitment of qualified researchers, such research cannot be accomplished.*

Establishing and maintaining high-quality and relevant data systems require a sufficient and capable intramural research staff that is diversified across disciplines. The current impoverished research capacity of SSA not only affects the timely analysis of data collected, but also leads to an inability to anticipate important issues and respond to them. *The intramural staff for disability research and statistics has to be substantially expanded and diversified to implement the recommendations in this report.*

Moreover a balanced program of intramural and extramural research is needed. An extramural research program, however, places its own demands on the agency's research staff. Oversight responsibility rests with the agency for careful evaluation of the work of the external researchers to ensure its quality, adequacy, and appropriateness. The committee also believes that a strong peer-reviewed extramural program is needed in the social insurance area. *SSA should expand and diversify its extramural research program to include a balance of contracts, cooperative agreements, and investigator-initiated grants.*

As this report makes abundantly clear, SSA has been given a difficult task and dwindling resources to deal with it. The situation will get worse, not better, in light of the anticipated growth in the demands on the program as the baby boom generation reaches the age of increased likelihood of disabilities. *Major rethinking of the disability program is required.* Little doubt exists that the current system is inadequate. *The fundamental problems of Social Security's disability decision process are not adequately reflected in the agency's research agenda.* If not corrected, this situation will impair the ability of SSA to meet its policy needs in the twenty-first century. Without sufficient resources, however, SSA cannot accomplish this forward-looking agenda.

The committee's key message in this report flowing from its contract mandate is that SSA desperately needs a long-term, systematic research program to address the growing demands on its disability programs and to provide the basis for improvements in the disability determination process. For many years, disability policy has been guided largely by court decisions and other pressures rather than by well-thought-out principles. No single source of policy has existed to which decision makers can turn for direction.

Although during the course of its study the committee identified much that needed changing, and continues to be concerned about some of the decisions made by SSA, it recognizes that SSA has made several modifications in response to its recommendations for improving the National Study of Health and Activity. *The committee believes that the blueprint for action that it recommends for developing and implementing a disability monitoring system for Social Security programs, and for needed research relating to improving the disability decision process, will contribute toward a significantly improved and efficient system of measuring and monitoring work disability that will better inform public policy and serve the public. This blueprint is worthy of full funding and adequate staffing support by both the Congress and the executive branch of the government.*

RECOMMENDATIONS

On the basis of its findings and conclusions the committee provides four categories of recommendations: conceptual issues in defining disability, survey measurement and monitoring of disability, improving the disability decision process, and enhancing research resources. The text of the committee's recommendations, grouped according to these categories follows, keyed to the chapter in which they appear in the body of the report.

Conceptual Issues

Recommendation 3-1: The committee recommends that the Social Security Administration develop systematic approaches to incorporate economic, social, and physical environmental factors in the disability determination process by conducting research on:
• the dynamic nature of disability;

- the relationship between the physical environment and social environment and work disability; and
- understanding the external factors affecting the development of work disability.

Survey Measurement and Monitoring of Disability

Recommendation 4-1: The committee recommends that prior to undertaking any future large-scale data collection effort, the Social Security Administration should allow for sufficient time and provide adequate resources to systematically:

a. investigate, test, and incorporate conceptual developments; and

b. develop, pretest, pilot and revise measurement instruments and design.

Recommendation 5-1: The committee recommends that the Social Security Administration develop an ongoing disability monitoring system building from its experience with the National Study of Health and Activity.

The committee further recommends that the Social Security Administration establish a clear set of objectives for guidance in developing and implementing the substantive content of the system.

Recommendation 5-2: The committee recommends that the disability monitoring system consist of

a. a periodic, comprehensive, and in-depth survey to measure work disability; and

b. a small set of core measures in the intervening years derived from surveys and, or, administrative data.

SSA should collaborate with other federal agencies on the design and implementation of the monitoring system.

Recommendation 5-3: The committee recommends that the Social Security Administration establish a continuing, external technical committee of experts for the planning and implementation of the recommended disability monitoring system.

Improving the Disability Decision Process

Recommendation 6-1: The committee recommends that prior to making the changes in the current decision process, SSA should
a. establish evaluative criteria for measuring the performance of the decision process;
b. conduct research studies and analyses to determine how the current processes work relative to these preestablished criteria; and
c. evaluate the extent to which change would lead to improvement.

Recommendation 6-2: The committee recommends that the Social Security Administration conduct research on
a. improving the ability to identify and measure job requirements for the purpose of determining work disability;
b. investigating the role and effects of vocational factors in the disability decision process; and
c. understanding reasons for variations in allowance rates among states and over time.

Recommendation 6-3: The committee recommends that the Social Security Administration initiate a research program for testing decision process models that emphasizes rehabilitation and return to work.

Enhancing Research Resources

Recommendation 7-1: The committee recommends that the intramural staff for disability research and statistics should be substantially expanded and diversified to implement the recommendations in this report.

Recommendation 7-2: The committee recommends that the Social Security Administration (SSA) expand and diversify its extramural research program to include a balance of contracts, cooperative agreements, and investigator-initiated grants. This broadened research program would prepare the SSA for the anticipated growth in the demands on the disability programs and to bring about the needed fundamental changes in its disability programs.

Part I

REVIEW AND RECOMMENDATIONS

1

Introduction

The Social Security Disability Insurance program (Title II of the Social Security Act (hereafter, "the Act") and the Supplemental Security Income program (Title XVI of the Act) are the two major federal programs providing cash benefits and eligibility for medical benefits to persons with disabilities. The Social Security Disability Insurance program (SSDI) is an insurance program that provides payments to persons with disabilities based on their having been covered previously under the Social Security program. The Supplemental Security Income (SSI) program is a means-tested income assistance program for disabled, blind, and aged persons who have limited income and resources regardless of their prior participation in the labor force.

The definition of disability and the process of determining disability are the same for both programs. The Social Security Act defines disability (for adults) as ". . . inability to engage in any substantial gainful activity by reason of any medically determinable physical or mental impairment which can be expected to result in death or which has lasted or expected to last for a continuous period of not less than 12 months . . ." (Section 223 [d][1]). Amendments to the Act in 1967 further specified that an individual's physical and mental impairment(s) must be ". . . of such severity that he is not only unable to do his previous work but cannot, considering his age, education, and work experience, engage in any other kind of substantial gainful work which exists in the national economy, regardless of whether such work exists in the immediate area in which he lives, or whether a specific job vacancy exists for him, or whether he would be hired if he applied for work" (Section 223 and 1614 of the Act). During the past two decades, SSDI and SSI programs have experienced faster than expected growth. In 2000 the Social Security Administration (SSA) paid $50 billion in cash benefits to 5.0 million workers under the SSDI program. Between 1989 and 2000, the number of workers receiving SSDI benefits rose from about 2.9 million to nearly 5.0 million, an increase of almost 74 percent. Likewise, in 2000, SSA paid $19 billion in benefits to 4.0 million blind and disabled working age people under the SSI program, an increase of 74 percent between 1989 and 2000 (SSA, 2001d). To a large extent this growth reflects the increases in the number of people applying for and entering the programs and a decrease in the number leaving the programs.

BACKGROUND

Statement of the Problem

Historically the disability program has been subject to rapid increases followed by periods of decline in rates of application, awards, and terminations. These fluctuations appear to arise both from external forces and from program and policy shifts. In the future, disability policymakers must have the ability to carefully gauge the effect of any policy changes in order to avoid excessive shifts in program experience resulting from such action that may stimulate, in turn, major policy reactions in the opposite direction. The challenge for SSA is to understand the reasons for fluctuations in the growth of disability rolls in order to better manage the programs and guide the anticipated growth over the coming decades.

In 1992, the Board of Trustees of the Old Age Survivors Insurance and Disability Insurance Trust Funds requested the Department of Health and Human Services (DHHS) to conduct an analysis of the SSDI program experience to explain the rapid program growth before the Board could make any recommendation to the Congress on statutory adjustments (DHHS, 1992). The DHHS study found that although the increases in applications for adult disability benefits cannot be explained definitively, many factors may have contributed to the growth in the number of people receiving Social Security disability benefits. These factors include the economic downturn in the late 1980s and early 1990s in the United States; structural changes in the labor market; demographic trends such as changes in the size, composition, and characteristics of the working age population; changes in public policies and the types of disabling impairments that are recognized and diagnosed for disability cash benefits; and a decrease in the average age of new beneficiaries with a resulting increase in the duration of benefits. Programmatic factors include increase in program outreach and public awareness, changes in other support programs, and cost shifting by states associated with cuts in state and locally funded general assistance and other welfare programs, and the deinstitutionalization of people with mental disorders and mental retardation and other disabilities who were previously cared for by state hospital systems.

The analysis further showed that over the years, legislative and regulatory changes and judicial interpretations of eligibility criteria also have extended the scope of the program. Many other factors also have contributed to the growth of the programs, such as incentives to apply for benefits affected by changes in the structure of alternative public and private income support programs for persons with disabilities and the increases in benefit amounts and level of substantial gainful activity allowed for people receiving disability benefits. (These factors and their impact on the growth of the disability programs are discussed further in Chapter 2.)

As a result, SSA often has been faced with large workload increases in the disability programs and consequent backlogs in processing claims and appeals. These increases, however, have not been matched by increases in administrative resources. This imbalance has resulted in significant delays in processing disability claims determinations. A study conducted by SSA (1993) of the disability claim and appeal processes found that the processing time for a claim from the initial inquiry through receiving an initial claims decision notice can take up to 155 days, and through receipt of hearing decision notice, can take as long as 550 days. However, the actual time during this period that employees devote to working directly on a claim was found to be 13 hours up to the initial decision notice and 32 hours through receipt of hearing decision notice. The need to develop extensive medical evidence in every case, delays in the receipt of required medical evidence and consultative examinations at each level, and the wait at each stage of the application process because of missing information as the case is developed, impede timely and efficient decision making (SSA, 1994a).

Errors in making denial decisions by the state Disability Determination Service (DDS) adjudicators, backlogs in appeals, and inconsistencies in decisions reached by DDS adjudicators and administrative law judges (ALJs) are also a matter of concern. The decision-making standards and procedures used by the ALJs are not always the same as those followed by the DDS adjudicators. The subjective element in the disability decision process also contributes to the differences in disability decisions made at different levels of the application process and among different states (DHHS, 1982; GAO, 1994, 1997b; Hu et al., 1997). The number of decisions being appealed for reconsideration and then approved at the higher level has increased. Over time the process has become lengthy and complicated, burdened by complex policies and procedures applied at different levels, resulting in untimely and inconsistent decisions (SSA, 1994a; GAO, 1995, 1997a; SSAB, 2001).

Despite all these factors and the resulting workload increases, the procedures in the current disability process have not changed in any major way since the beginning of the SSDI program in the 1950s.

Origins of the Committee's Study

On the basis of its analysis, DHHS concluded that to better understand the need for disability benefits in the 1990s and beyond, a survey of health and disability in the United States, similar to such surveys completed in the late 1960s and the 1970s, should be undertaken. Such a survey also could assist in estimating the future cost of the disability program. Based on these findings and conclusion, the Board of Trustees recommended that the DHHS initiate a significant research effort to establish more clearly whether the SSDI program's rapid growth in the 1990s was a temporary or a longer-term phenomenon.

In response to this recommendation, SSA initiated research aimed at understanding the growth of disability benefit programs—the changes in the size of the potentially eligible population, changes in the behavior of potential beneficiaries with respect to applying for benefits, changes in award rates, and the length of time beneficiaries remain on the rolls (DHHS, 1992; Muller and Wheeler, 1995). A number of research projects were initiated, including staff analyses and contracts to undertake econometric analyses of the causes of disability growth using cross-sectional data; a survey of field office managers undertaken by SSA as part of the research effort to understand the changes in the application behavior of individuals who are potentially eligible for disability benefits; and the Disability Evaluation Study (DES), later renamed by SSA the National Study of Health and Activity (NSHA). SSA views NSHA as the cornerstone of its long-term disability research agenda to improve its ability to understand the growth of the disability programs and to estimate the current and projected pool of the eligible population, the number who may apply for benefits, and the number who may be awarded benefits and their characteristics. It is a complex multiyear national survey of the United States household population 18–69 years of age.

Concerns about the numerous long-standing problems and complaints relating to the disability determination process, summarized above, led the SSA leadership to fundamentally rethink the entire process for determining program eligibility and improve the quality of the service in the disability claims process. In the early 1990s, the National Performance Review, headed by Vice President Gore, also directed improvement of the

Social Security Administration's disability process as one of the key service initiatives for the federal government (SSA, 1994a). SSA believed that significant improvements could not be achieved without fundamentally restructuring the entire claims process. In view of these numerous concerns and the agency's recognition of the need to improve the quality of the service in the disability claims process, SSA decided to develop an ambitious long-term strategy for reengineering "… the disability determination process that would be simpler than the existing one, deliver significantly improved service to the public, remain neutral with respect to program dollar outlays, and will be more efficient to administer" (SSA, 1994a, p. 46). It further stated that ". . . unless SSA invests substantially more funds to research and development of the simplified disability determination methodology, the full benefits of the redesigned process . . . will not be possible" (p. 46).

DISABILITY DETERMINATION—STRUCTURE AND PROCESS

Disability Claims Process [1]

The Social Security disability claims process starts at the state Disability Determination Service where most disability decisions are made for SSA at the initial and reconsideration levels. Briefly, the claims process proceeds through a series of four stages or levels: (1) applications for benefits and preliminary screening are made at the SSA district offices; (2) disability determinations are made in state DDS agencies using federal regulations and SSA guidelines and procedures; (3) claimants whose applications are denied can have their claims reconsidered at the DDS level; and (4) if benefits are denied during the reconsideration, the claimant may request a hearing before an ALJ at the SSA. Further appeals options include a request for review of the denial decision by SSA's Appeals Council, and then review in the federal courts.

SSA envisioned that the reengineered claims process would make efficient use of technology, eliminate fragmentation and duplication, and promote flexible use of resources. Claimants would be given understandable program information and a range of choices for filing a claim and interacting with SSA. They would deal with one contact point and would have the right to a personal interview at each level of the process. Also, the number of levels in the new claims process prior to Appeals Council

---

[1]For a more detailed description of SSA's claims process and its plans for reengineering, the reader is referred to *Plan for a New Disability Claim Process* (SSA, 1994a) and *Disability Process Redesign: Next Steps in Implementation* (SSA, 1994b).

review would be consolidated from four to two, and the issues for which appeals would be allowed would be more focused. Finally, if the claim is approved, the initiation of payment would be streamlined (see Figure 1-1).

Successful reengineering depends on a number of key initiatives of a new claims process. SSA's original plan depended on a large number of initiatives that together were intended to make the reengineered claims process function efficiently. Since then the agency has reassessed many of the reengineering initiatives and developed a revised plan that focused on eight major areas for priority attention. Four of these initiatives are testing efforts (single decision maker, adjudication officer, full process model, and disability claims manager), and four are developmental activities that SSA calls "critical enablers" (systems support, process unification, simplified decision process, and quality assurance) (SSA, 1998). Thus the redesign of the disability decision process is only one of the process changes proposed by SSA to achieve reengineering of the disability claims process.

Evaluation of Eligibility for Disability Benefits

The Current Decision Process for Initial Claims

The disability decision[2] process for initial claims involves five sequential decision steps (SSA, 1994a).

1. In the first step, or point of decision, the SSA field office reviews the application and screens out claimants who are engaged in substantial gainful activity (SGA).[3]
2. If the claimant is not engaged in SGA, step two determines if the claimant has a medically determinable severe physical or mental impairment. The regulations define severe impairment as one that significantly limits a person's physical or mental ability to do basic work activities.
3. The documented medical evidence is assessed against the medical criteria to determine whether the claimant's impairment meets or equals the degree of severity specified in SSA's "Listings of Impair-

---

[2]Throughout the report the terms "disability decision" and "disability determination" are used interchangeably.
[3]In 2002 the SGA earnings level for nonblind beneficiaries is $780 a month (net of impairment-related work expenses), based on regulations published by the Commissioner in December 2000. These regulations provide automatic yearly indexing of the SGA monthly amount.

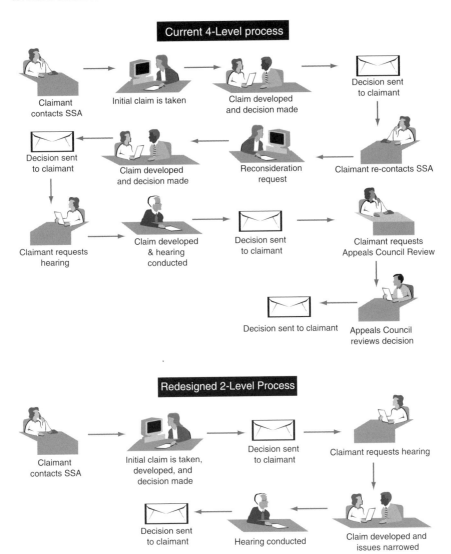

FIGURE 1-1 The Social Security Administration's current and proposed disability claims process.
SOURCE: Adapted from Social Security Administration, 1994a.

ments" (Listings).[4] A claimant whose impairment(s) meets or equals those found in the Listings is allowed benefits at this stage on the basis of the medical criteria.

4. Claimants who have impairments that are severe, but not severe enough to meet or equal those in the Listings, are evaluated to determine if the person has residual functional capacity (RFC)[5] to perform past relevant work. Assessment of the RFC requires consideration of both exertional and nonexertional impairments. If a claimant is determined to be capable of performing past relevant work, the claim is denied.

5. The fifth and final decision step considers the claimant's RFC in conjunction with his or her age, education, training, and work experience, commonly referred to as vocational factors, to determine if the person can perform other work that exists in significant numbers in the national economy.

Proposed Redesigned Decision Process

As stated above, the redesign of the disability decision process is only one of the many process changes proposed in the reengineered disability claims process. SSA has stated that such a redesigned decision process should

- be simple to administer;
- facilitate consistent application of rules at each decision level;
- provide accurate and timely decisions; and
- be perceived by the public as straightforward, understandable, and fair.

As envisioned by SSA, the goal of the new process was ". . . to focus the new decision-making approach on the functional consequences of an

---

[4]The Listings of Impairments for each body system describe impairments that are considered severe enough to prevent an adult from doing any gainful activity or to cause marked and severe functional limitations in a child younger than 18 years of age. Most of the listed impairments are permanent or expected to result in death, but some include a specific statement of duration. SSA first included the Listings in its regulations in 1968 to help expedite the processing of disability claims under the SSDI program and for SSI since it began in 1974. For a detailed description of the Listings the reader is referred to SSA's publication *Disability Evaluation Under Social Security* (2001a).

[5]Residual functional capacity is defined by SSA as what the claimant can still do in a work setting despite the physical or mental limitation caused by his or her impairment(s) (20 C.F.R. 404.15).

individual's medically determinable impairment(s)" (SSA, 1994a, p. 21). According to SSA, in the proposed redesigned disability decision process the presence of a medically determinable impairment will remain a necessary requirement for eligibility, as required by the current law. The proposed redesigned process, however, would focus directly on developing new ways to assess the applicant's functional ability or inability to work as a consequence of the medical impairment and to rely on these standardized functional measures to reach decisions. Medical and technological advances and societal perceptions about work capacity of a person with disabilities appear to support a shift in emphasis from the current focus on disease conditions and medical impairments to that of functional inability. For example, people with disabilities are able to function with personal assistants and assistive devices.

The redesigned disability decision process, as conceived by SSA, involved four sequential steps for deciding if a claimant meets the definition of disability as defined in the Act.

1. The first step is the same as in the current process. It involves screening out applicants who are engaged in substantial gainful activity.
2. If the claimant is not engaged in SGA, the second step would evaluate if the applicant has a documented medically determinable physical or mental impairment. Under the proposed revision, however, a threshold "severity" requirement was no longer needed.
3. The third step would assess if the person's impairment is included in an index of disabling impairments (to be developed). The index would replace the current listings of impairments. It would contain a short list of impairments of such severity that, when documented, they could be presumed to result in loss of the person's functional ability to perform substantial gainful activity without the need to further measure the individual's functional capacity and without reference to the person's age, education, and previous work experience.
4. If the claimant's medical impairment(s) was not in the index, the fourth and final decision step would evaluate if the individual has the functional ability to perform any substantial gainful activity. These individualized assessments of functional ability would also take into consideration the effects of the vocational factors in determining the demands of the individual's previous work. Functional assessment instruments would be designed to measure an individual's abilities to perform a baseline of occupational demands that include the primary dimensions of work and that exist in significant numbers in the national economy (SSA, 1994a).

The final decision step of the proposed decision process subsumed both steps four and five of the current decision process. According to SSA, this step reflects the most significant change from the current decision process. SSA assumed that under this proposed decision process, the majority of claimants would be evaluated at this point using a standardized approach to measuring functional ability to perform work. Conceptually, standardized measures of functional ability that are universally acceptable would facilitate consistent decisions regardless of the professional training of the decision makers in the decision process.

After reviewing the proposed plan for the redesign of the disability decision process, the Commissioner of SSA, Dr. Chater, concluded that those aspects of the proposal that deal with functional assessment, baseline of work, and the evaluation of age would require extensive research, testing, and deliberation with experts and consumers to determine whether they could be implemented (SSA, 1994a).

The current and proposed disability decision processes and the related research conducted by SSA are discussed in more detail in Chapter 6 of this report.

THE NATIONAL ACADEMIES STUDY

The Committee's Charge

In response to the Commissioner's directive, SSA launched a multiyear research effort to develop and test the feasibility, validity, reliability, and practicality of a redesigned disability determination process. SSA developed what it referred to as the *research plan* for the redesign of the disability decision process and a time line for its completion (SSA, 1996, 1997). In 1996, SSA requested that the Institute of Medicine (IOM), in collaboration with the Committee on National Statistics (CNSTAT) of the Division on Behavioral and Social Sciences and Education (DBSSE) of the National Research Council, conduct an independent, objective review of, and make recommendations on, the statistical design and content of the NSHA and on the approach, scientific method, adequacy, and appropriateness of the research plan for the redesign of the disability decision process. The study focuses on the working age population. The committee's specific tasks include, but are not limited to, the following:

- review the scope of work for the NSHA, request for proposal, and the design and content of the survey as proposed by the survey contractor;
- review and evaluate the preliminary design of the NSHA (the protocol developed by Westat), and subsequent modifications made

by SSA, identifying statistical design, methodological and content concerns, and other outstanding issues, and making recommendations as appropriate;

- review SSA's research plan and time line for developing a new decision process for disability and offer comments and recommendations on direction to the research; and
- review all completed research including, but not limited to, reviewing research into existing functional assessment instruments conducted under contract to SSA by Virginia Commonwealth University, and providing advice and recommendations for adopting or developing functional assessment instruments or protocols for the redesigned disability process and NSHA.

The IOM, in collaboration with CNSTAT, appointed a committee of 14 members representing a range of expertise related to the scope of the study. The committee held its first meeting in January 1997.

Realizing that some of the key components of the research and testing relating to the NSHA design will not be completed on schedule, SSA extended the contract period of four years for an additional two years to ensure the committee's review and evaluation of the results of the pilot study and any consequent proposed changes in the design and instruments for the national survey and other outstanding issues. In late 1999, SSA informed the committee that it had decided to no longer actively pursue the development of the new decision process as proposed in its disability redesign plan, instead it would focus its attention at this time on making improvements within the current decision process (SSA, 1999b).

Study Method

The committee executed its charge through the conduct of several activities. It reviewed and analyzed an extensive body of research literature, published and unpublished, and other documents including planning documents, internal papers, requests for proposals, relevant internal documents and unpublished papers related to the redesign and SSA's research plan, and other material provided by SSA and other government officials during the course of the study, as well as historical documents and publications relating to the subjects under consideration. Published literature on survey design and methods, evaluation and research on labor market trends, disabilities caused by physical and mental impairments, functional measures, and other relevant topics also were reviewed.

The committee met on 12 separate occasions between early 1997 and January 2002 to deliberate on the issues outlined above. Experts were invited to address the committee on various issues at five of these meetings. A listing of the committee meetings and the presenters can be found in Appendix A. Several subcommittee meetings were held to work on specific issues.

The committee held two large workshops to augment its knowledge and expertise by more focused discussion on specific issues of concern, and to obtain input from a wide range of researchers and other interested members of the public:

- The Workshop on Functional Capacity and Work was held on June 4–5, 1998.
- The Workshop on Survey Measurement of Work Disability was held on May 27–28, 1999.

The agendas, presenters, and discussants for both of these workshops can be found in Appendix B. Reports of the workshop deliberations were published (IOM, 1999a, 2000).

In order to provide timely advice to SSA as it developed its research and its survey method and content, the committee issued fast-track interim reports on specific targeted topics that needed immediate attention by SSA. Three such reports were issued:

- *Disability Evaluation Study Design.* First Interim Report (IOM, 1997)
- *The Social Security Administration's Disability Decision Process: A Framework for Research.* Second Interim Report (IOM, 1998)
- *Review of the Disability Evaluation Study Design.* Third Interim Report (IOM, 1999b)

The first interim report was limited to an examination of the general features of the proposed survey design, data collection plans, coverage, and sampling as described in the scope of work dated July 30, 1996 (SSA, 1996), in the draft request for proposals (RFP) developed by SSA for a contract to conduct the survey. The committee made no attempt in that report to comment on the content of the questionnaires, specific measures of functional capability, or the content of the medical examinations and medical and diagnostic tests proposed for the survey.

The second interim report was a preliminary assessment of the adequacy of SSA's research plan for developing a new disability decision process and the time line for its completion. In that context, the report outlined a framework for a research design and reviewed the general features and directions specified by SSA in the scope of work in the relevant requests for proposals for the conduct of the research. It identified critical elements of a research design that were missing from SSA's current plans, and offered suggestions for changes in priorities and improvements in the research projects already under way and others yet to be developed.

The third interim report was related directly to one of the contract tasks—review of the design, approach, and content of the survey, as proposed by SSA's contractor for the survey, Westat, Inc. The report was a brief review of sample design, including that of the pilot study, instruments and procedures, and response rates goals developed by the survey contractor, Westat, and provided to the committee by SSA in June 1999 for its review and recommendations (Westat, 1999a,b,c). The report also commented on the proposed time line for initiation of each phase of the survey.

Detailed technical recommendations made by the committee in these interim reports are listed in Appendix C. The recommendations in the final report build on those of the interim reports.

To avail itself of expert and detailed analysis of some of the key issues beyond the time and resources of its members, the committee commissioned five background papers listed below from experts in areas of concept and measurement of disability, survey design and method, mental impairments, and disability and the labor market:

1. "Conceptual Issues in the Measurement of Work Disability," by Alan Jette, Ph.D. and Elizabeth Badley, M.D.
2. "Methodological Issues in the Measurement of Work Disability," by Nancy Mathiowetz, Ph.D.
3. "SSA's Disability Determination of Mental Impairments: A Review Toward an Agenda for Research," by Cille Kennedy, Ph.D.
4. "Survey Design Options for the Measurement of Persons with Work Disabilities," by Nancy Mathiowetz, Ph.D.
5. "Persons with Disabilities and Demands of the Contemporary Labor Market," by Edward Yelin, Ph.D., and Laura Trupin, MPH.

The full text of these papers is included in Part II of this report.

SCOPE AND LIMITATIONS

The scope of the present study is broad, consisting of two components: (1) ongoing detailed review and advice on the design, methods, sampling and content of a major complex survey of disability, and (2) a review of the research plan and the individual research projects undertaken by SSA to guide it in redesigning the disability decision process. The statistical design, methods, and content of the NSHA and the research plan for the redesign of the disability decision process represent two separate subject areas of study, each with different issues. For the most part, therefore, they are discussed separately in this report. The study component relating to the redesign initiative is limited to the review and advice on the research being conducted and planned for the redesign of the disability *decision* process, which is only one element of SSA's total effort to reengineer the disability *claims* process.[6] The scope and extent of the review of survey plans, as well as the redesign research plans and individual research projects, were dependent on what was initiated or completed and made available by SSA to the committee during the course of the study.

Defining Work Disability

Agreement does not exist on how to define and measure disability (Frey, 1984; Kennedy and Gruenberg, 1987; Verbrugge, 1990; Mather, 1993). There is ongoing debate about the general concept of disability, some of which is discussed in Chapter 3. SSA's focus in both the SSDI and the SSI programs, is on *work disability* as defined by the Social Security Act. As stated earlier in the chapter, the Act defines disability (for adults) as inability to engage in any substantial gainful activity anywhere in the national economy by reason of any medically determinable physical or mental impairment that can be expected to result in death or that has lasted or is expected to last for a continuous period of not less than 12 months. An individual's physical and mental impairment(s) must be of such severity that he or she not only is unable to do the previous work but cannot, given the person's age, education, and work experience, engage in any other kind of substantial gainful work that exists in the national economy, regardless of whether such work exists in the immediate area in which the person lives, or whether a specific job vacancy exists, or whether they would be hired if they applied for work. The definition makes clear that these programs deal with *work disability*.

In recent years the concept of disability has generally shifted from a focus on diseases, conditions, and impairments to one on functional limitations caused by these factors (Adler, 1996). SSA's definition of disability was developed in the mid-1950s at a time when a greater proportion of jobs were in manufacturing and required physical labor than is the situation today. It was therefore expected that people with severe impairments

---

[6] SSA's reengineering plan focused on eight major areas for priority attention. Four of these initiatives were testing efforts (single decsion maker, adjudication officer, full process model, and disability claims manager), and four were developmental activities that SSA calls "critical enablers" (system support, process unification, simplified decision process, and quality assurance) (SSA, 1998).

would not be able to engage in substantial gainful activity. Over the years, many changes have occurred. As the nature of work has shifted from the manufacturing to the service sector, more severely disabled persons are able to be employed because of medical and technological advances; and in recent years the public's attitude about the employment of people with disabilities also has changed as reflected in the Americans with Disabilities Act of 1990 (ADA). In light of these changes, critics claim that SSA's process of determining disability has not kept pace either with the understanding of disability or with advances in medical science and changes in the organization of work.

ORGANIZATION OF THE REPORT

The committee used three criteria for judging the contents of this report and its specific recommendations. First, the topic examined should be relevant to and within the scope and purview of the committee's charge. Second, the evidence and analysis must be sufficient to support and justify its findings and recommendations. Third, a recommendation should be attainable at reasonable cost.

The research plan for the redesign of the disability decision process and the scope, statistical design and methods, and content of the NSHA represent separate, and yet related, subject areas of study with different issues. For the most part, therefore, they are discussed separately in the report. The report summarizes as appropriate the key conclusions and recommendations made by the committee in its interim reports to SSA during the course of the study and discusses the need for, and makes recommendations for, the development and maintenance of a national system to monitor the disability programs on an ongoing basis and the conduct of research needed to improve its evaluation of eligibility for disability benefits. This report is organized in a manner responsive to the contract charge.

Chapter 2 provides a brief overview of disability trends and discusses some of the factors that may have contributed to these trends.

Chapter 3 describes the meaning of the term *disability* and the relationship between the generic concept of disability and the term *work disability*.

Chapter 4 briefly reviews the design, sample size, content, and time line of the NSHA. The chapter then discusses continuing issues in survey measurement of disability and work disability, relating them to problems encountered in the research development, design, and time line of the NSHA. Finally it lays out a program of research in survey measurement issues that need to be addressed by SSA, other federal agencies, and other researchers and makes recommendations relating future surveys of disability and work.

Chapter 5 explores ways in which SSA could build on its experience with the NSHA to develop an ongoing disability monitoring system for Social Security programs that would provide timely information on the prevalence of disability and the characteristics and distribution of persons with disabilities. The chapter discusses the need for and elements of such a system, a brief description of possible survey partners in the development and use of the data, the essential principles for such a system, a needed advisory structure, and a suggested development and implementation strategy.

Chapter 6 summarizes the committee's preliminary assessment undertaken early in the study of SSA's research plan to redesign the disability decision process (IOM, 1998), and the subsequent decision by SSA to terminate this redesign effort and explore ways to incrementally improve the current process. It makes recommendations on research needed to improve the disability decision process.

The final chapter highlights some of the broad issues of analytical capacity and resource considerations to implement the recommendations embodied in this report. The chapter closes with a call for needed research that would lead to fundamental improvement in the research and administrative structure and policy in the disability programs.

Although this report addresses the specific tasks in the committee's mandate—to review the research related to the redesign of the disability decision process and the design, scope of work, and content of the NSHA—the committee hopes that the report will provide guidance to a wider audience responsible for disability policy and to researchers concerned about enhancing the ability to measure disability in a survey context. Further, the report should contribute toward development of an efficient and cost-effective system for ongoing monitoring of the prevalence of disability in the United States to guide the future direction of disability policy.

2

Dynamics of SSA's Disability Programs

The dynamics of the disability programs have been shaped over the years by many events. Economic conditions, demographic changes, public opinion, and resulting congressional and Administrative actions have had a significant impact on program experience. This chapter reviews the historical development and growth of the disability programs administered by the Social Security Administration (SSA) for the working age population in the 45 years since the inception of the Social Security Disability Insurance (SSDI) program in 1956 and in the 30 years since the Supplemental Security Income (SSI) program in 1972. The chapter further discusses some of the main factors—economic and noneconomic, intrinsic and extrinsic to the program—that have shaped these programs over the years. These include legislative initiatives and judicial decisions, the demographic composition and characteristics of the population, the types of impairments of applicants, incentives and outreach, and the changing nature of work.

Although the decision of an individual to apply is an important variable in the program size, the program's eligibility requirements affect its ultimate size. Moreover, the stringency or leniency of program implementation impacts the size and cost of the program and also the probability of a person's applying. Growth in the initial awards (or allowances) often is attributable to some of the same factors that are associated with growth in applications during that period. A better understanding of the dynamics of the disability programs is essential to enhance the ability to predict the future growth and cost of the program.

In the 1970s when the disability programs were growing rapidly, econometric research studies using aggregate time series techniques were undertaken to understand the role of the various factors in this growth. Recent rapid growth in the programs again has focused attention on the need to undertake a rigorous research program to estimate the extent of disability in the United States and to determine the potential need for disability benefits in the twenty-first century. Unfortunately, analysts have been faced with a paucity of current information since the late 1960s and 1970s when SSA conducted three surveys of disability and work. These surveys obtained information on impairments and various socioeconomic factors that were useful in the analysis of disability programs. Likewise, any legislative and/or administrative initiatives to increase control over the program size and to improve the processing of claims should be based on research aimed at understanding the relative roles of the various variables that impact on disability programs administered by the SSA. Some of these factors may be within the control of the Congress and the Administration, while others may be outside their purview.

HISTORICAL DEVELOPMENTS AND PROGRAM GROWTH[1,2]

The need for disability insurance was recognized in the late 1930s when the Social Security Act was enacted. For many years, Congress and the Administration were hesitant to enact such a program because of concerns about the difficulties in deciding whether a particular person is disabled and in containing costs and predicting future program growth. These concerns have remained to the present day.

However, a Social Security Disability Insurance program was enacted in 1956 to provide cash benefits to a person unable to engage in substantial gainful activity (SGA) ($780.00 per month in 2002) by virtue of a medical impairment that was expected to result in death or be of long-continued or indefinite duration. The Act gave states responsibility for initial disability determination, acting under contract with the federal government. Reflecting the concerns about containing costs, it limited disability benefits to individuals 50–64 years old and did not extend benefits to the dependents of disability beneficiaries. A separate disability insurance tax rate and trust fund were established to allow close monitor-

---

[1]Much of the information in this section is excerpted from DHHS, 1992; Berkowitz, 1997; and Mashaw, 1997. The statistics presented are mostly published data from the Social Security Administration.

[2]Consistent with the mandate of the study, the discussion and statistics presented in this chapter for the most part relate to disabled workers for the SSDI program and the working age population (18–64 years of age) for SSI program.

ing of program costs. The first payments were made in 1957. At that time, SSDI was thought of as a source of early retirement benefits mainly for men who had worked most of their lives but became disabled with chronic diseases of aging close to the normal age of retirement.

Period of Growth

Over the years the SSDI program has steadily, if not uniformly, expanded its coverage and support levels. The program grew rapidly in the early 1960s and through the middle of the 1970s. Several amendments to the Social Security Act extended the qualifying requirements for disability benefits. In 1958, benefits were extended to dependents of beneficiaries. The 1960 amendments extended benefits to all qualified persons under 65 years of age. These alterations changed the concept of SSDI from being an alternative to retirement to an alternative to working. The legislative amendments of 1965 made the definition of disability more liberal by requiring only that the impairment be expected to result in death or to last for at least 12 months. The 1967 amendments eased the insured status requirements for persons under age 31, allowing a substantial number of young beneficiaries to enter the rolls. These amendments led to an increasing proportion of younger and relatively healthier beneficiaries. The required waiting period before receiving benefits was reduced from six to five months in 1972. The level of SSDI benefit amounts was increased in the early 1970s and automatic cost-of-living adjustments were enacted. Also in 1972, Medicare coverage was extended to persons who had received disability benefits for two years. By the mid-1970s these changes had resulted in higher replacement rates of prior earnings, making it more financially attractive for people to apply for benefits and for beneficiaries to remain on the rolls.

These changes defined a much larger pool of persons potentially qualified for entitlement. The early 1970s experienced a rapid increase in the number of applications and awards. During the period 1960–1975, the number of applications grew rapidly from about 418,000 in 1960 to nearly 1.3 million in 1975. During the same period the number of awards grew from about 200,000 in 1960 to almost 600,000 in 1975. Figure 2-1 shows the number of applications, awards, beneficiaries on the rolls, and terminations of disability worker benefits from 1960 to 2000. In relative terms, the number of applicants grew from 8.6 to 15 per 1,000 persons insured in case of disability from 1960 to 1975 (see Table 2-1 below).

In 1970, Congress enacted legislation establishing the Black Lung Program and in 1972 enacted legislation establishing the Supplemental Security Income program for the aged, blind, and disabled (P.L. 92-603). These programs, especially SSI, had a major impact on the growth and manage-

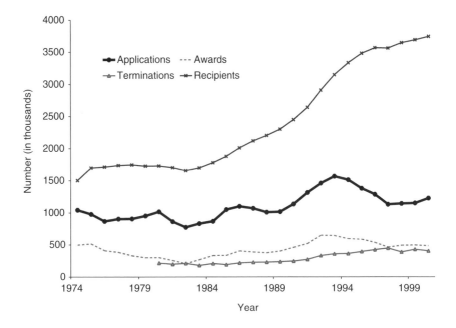

FIGURE 2-2 Number of SSI applications, awards, recipients, and terminations, aged 18–64 years, 1974–2000.
SOURCE: SSA, 2001b. 2001c.

Processing this large workload with limited staff resources led to administrative expediencies in the requirements for processing disability claims. Central Office reviews of DDS decisions for quality assurance fell from 100 percent to about 5 percent in 1972, and they were conducted after, rather than before, payment of benefits began. Most reviews were deferred, and only obvious decision errors were returned for correction.

The legislative changes, increased benefit levels, changes in program administration, and the economic downturn of the early 1970s probably contributed to the sharp increase in the disability incidence rate (number of new SSDI benefit awards per 1,000 workers insured in case of disability) from 4.8 in 1970 to a high of 7.1 in 1975 (Table 2-1). At the same time the termination rate (proportion of beneficiaries whose benefits were terminated) declined from 174 per 1,000 beneficiaries in 1970 to 132 per 1,000 in 1975 (Table 2-2). Terminations of SSDI benefits occur as a result of death, conversion of disability benefits to old age and survivor benefits upon attainment of normal retirement age (currently 65), or recovery (beneficiary no longer meets the standards used to define disability, either

TABLE 2-2 Disabled Workers: Number of SSDI Beneficiaries and Terminations and Termination Rate per 1,000 Beneficiaries, 1960–2000

Year	No. of Beneficiaries (thousands)	No. of Terminations (thousands)	No. of Terminations per 1,000 Beneficiaries
1960	455.4	89.1	195.6
1965	988.1	156.6	158.5
1970	1,492.9	260.4	174.5
1975	2,488.8	329.5	132.4
1980	2,858.7	408.1	142.7
1985	2,656.6	340.0	128.0
1990	3,011.3	348.2	115.6
1995	4,185.3	399.5	95.4
2000	5,042.3	460.4	91.3

SOURCE: SSA, 2001d.

because of medical recovery or return to work).[3] More than half of the decline from 1970 to 1975 in termination rates was due to lower medical recovery rates.

Period of Retrenchment

The rapid growth in disability rolls during this period renewed past concerns about SSA's ability to control program growth and the unpredictability of program expenditures. This situation led to legislative and administrative changes in the program in the late 1970s and early 1980s, slowing the increase in disability program growth.

The legislative amendments in 1977 reduced the income replacement rates in SSDI, particularly for younger beneficiaries. The 1980 amendments mandated reviews of 65 percent of allowed claims in the ensuing three years prior to the start of any payment. They also required a vastly increased process of review of the eligibility of existing disability beneficiaries. The 1980 amendments also limited the total amount of monthly Social Security benefits that could be paid to a disabled worker and his family by enacting replacement rate caps, further modified the calculation of benefits for younger disabled workers, and added work incentives. In 1981, Congress eliminated the minimum Social Security benefit for new beneficiaries (DHHS, 1992; Mashaw, 1997).

---

[3]For a detailed discussion of the experience of disability benefit terminations, the reader is referred to *Social Security Disability Insurance Program Worker Experience* (SSA, 1999c).

SSA also refined its regulations and guidelines, changed the instructions and training to state Disability Determination Services to make eligibility criteria and evidentiary requirements more stringent, made quality assurance reviews more stringent, and increased the number of continuing disability reviews (CDRs).

These actions had a dramatic impact on applications for benefits and initial award decisions. The proportion of claims awarded benefits by the DDSs declined from 46 percent of the claims in 1975 to 31 percent in 1980 (see Table 2-1), and terminations rose to almost 143 per 1,000 beneficiaries by 1980 (see Table 2-2). The rate of applications also declined from 15 per 1,000 insured workers in 1975 to 12.6 in 1980. Although the economy was in decline, the number of new awards dropped sharply and the number of persons discontinued for "medical and return-to-work recovery" reasons increased (SSA, 2001d).

Period of Slow Growth

These legislative and administrative changes and resulting practices faced strong resistance both in the courts and in state governments, and led to widespread criticism in the media. Negative publicity over the large numbers of beneficiaries—particularly the mentally impaired— being removed from the disability rolls led to a reconsideration of the changes in disability programs.

By 1984, another reversal in attitudes occurred followed by another round of legislative and administrative changes. SSA placed an administrative moratorium on the conduct of CDRs. A series of congressional hearings were held highlighting the plight of beneficiaries removed from the rolls. Several legislative and judiciary actions undid many of the stringent policies that had produced the retrenchment during the late 1970s and early 1980s. Court cases and class action suits increased dramatically, and many persons were returned to the rolls through court appeals. Congress enacted the Social Security Benefits Reform Act of 1984. Its provisions included more liberal standards for mental impairments that emphasized the individual's ability to perform substantial gainful work, consideration of combined effects of multiple impairments in the absence of a single severe impairment, requirement for proof of medical improvements before termination of benefits, and use of SSA's regulatory standards to evaluate the effect of pain on disability. Court decisions on class action suits during the middle 1980s resulted in placing more emphasis on the opinion of treating physicians in the disability determination process, the role of pain as a disabling factor, and evaluation of a person's functional limitations in addition to the medical condition.

Once again, both applications and awards began to rise. The initial allowance rate that had declined to a low of around 31 percent of applications in 1980 and 1981, increased steadily during 1985–1989 and remained at about 35–44 percent during 1985–1990 (see Table 2-1). The final implementing regulations revising the eligibility criteria for mental impairments were published in 1986, resulting in dramatic increases in the number of benefits awarded on this basis. The number of awards to individuals with disabilities based on AIDS or HIV infection contributed to this increase. The termination rates also declined significantly as a result of SSA's moratorium on CDRs and their subsequent reinstatement under new and less stringent standards (see Table 2-2).

During the latter half of the 1980s, after the brief increase in the late 1980s associated with adjudicating a large number of cases under the new regulations for mental impairments, applications and incidence rates for disability benefits remained fairly stable.

Growth in the 1990s

Although the legislative and administrative climate was relatively stable after 1985, applications and awards for disability benefits once again began to climb rapidly in 1989 and into the 1990s. Most of the increase in awards followed the sharp increase in applications for benefits accompanied by a small increase in the initial allowance rates. The economic downturn in 1990 and 1991 may account for part of this increase. Applications for SSDI benefits rose by 8.4 percent in 1990 over the previous year followed by another 13 percent increase in 1991. This growth resulted in an increase in the incidence rate from 3.7 per 1,000 in 1989 to 4.5 in 1991, a 21.6 percent increase over the two-year period (SSA, 2001d).

In recent years, well in excess of a million disabled workers have applied for SSDI benefits each year reaching 1.3 million in 2000. More than 600,000 disabled workers were awarded benefits in 2000. In contrast, the number of persons whose benefits have been terminated was around 460,000 in that year (see Figure 2-1). With the exception of the late 1970s and early 1980s, the proportion of SSDI beneficiaries whose benefits have been terminated has declined steadily from the earliest years of the program, from 132 per 1,000 beneficiaries in 1975 to nearly 143 per 1,000 in 1980, to 115.6 in 1990, and to about 91 per 1,000 in 2000 (see Table 2-2).

As shown in Table 2-3, with the exception of the period in the early 1980s, the overall number of beneficiaries on the rolls, as well as the rate per 1,000 persons insured in the event of disability, has increased steadily over time as the growth in awards has outpaced terminations. Most terminations occur as a result of death or conversion. The trend in terminations has been declining. Two significant factors contribute to this trend—

TABLE 2-3 Disabled Workers: Number of SSDI Beneficiaries, Workers Insured in Event of Disability, and Beneficiaries per 1,000 Insured, 1960–2000

Year	No. of Beneficiaries (millions)	No. of Workers Insured (millions)	No. of Beneficiaries per 1,000 Insured
1960	0.455	48.5	9.38
1965	0.988	55.0	17.96
1970	1.493	74.5	20.04
1975	2.489	85.3	29.18
1980	2.857	100.3	28.48
1985	2.657	109.6	24.24
1990	3.012	120.1	25.08
1995	4.185	128.2	32.64
2000	5.042	138.7	36.35

SOURCE: SSA, 2001d.

lower death rates as a result of people living longer and a reduction in the average age of beneficiaries.

The change in the number of persons 18–64 awarded SSI disability benefits and the total number of recipients over time is similar to the dynamics observed in the SSDI program. The rapid increase in the total number of SSI participants in the early 1990s is a function of the growth in the number of disabled persons among SSI applicants and the poor economy as the 1990s began. The growth in the number of disabled adults is complicated and not fully understood. The reforms of the early 1980s and the outreach efforts in the 1980s also resulted in increases in the SSI program. With the strong economy of the late 1990s, a modest decline in SSI program participation was noted. However, because relatively few persons leave the SSI rolls, the total number of recipients has risen steadily since the 1980s, with the exception of a slight decline in the late 1990s.

FACTORS CONTRIBUTING TO RECENT GROWTH

As stated above, applications and awards for disability benefits in both the SSDI and the SSI programs increased significantly in 1989 and into the 1990s. The reasons for this recent increase are complex and are not fully understood. A combination of many factors may have contributed to this growth—some may be related to the broader socioeconomic and demographic environment and others may be associated with programmatic actions and court decisions. Many of the same factors have had a role in the programs' growth since their inception and are contributing also to the recent growth of the disability programs of SSA. Some of these factors are discussed briefly below.

Demographic Trends

The number of persons who apply for and receive benefits is influenced by the size, composition, and characteristics of the potentially eligible population. The composition of the SSDI and SSI populations has changed dramatically since the programs' inceptions. The size of the insured population for disability insurance has grown primarily because the working age population has grown (and an increasing number of women have entered the labor force). Between 1980 and 2000 the population of workers 20–64 years of age insured in the event of disability grew from 56.6 million to 71.6 million for men and from 37.4 million to 62.5 million for women (SSA, 2001d). The working age eligible population is projected to increase in the coming years as the baby boom generation ages and reaches 40–50 years of age, when chronic disease and disabilities are more likely to occur.

The composition of the SSI population also has undergone a fundamental change since the program began in 1974. In the early years, nearly 60 percent of the recipients were aged. Over the years, the number of aged beneficiaries has declined significantly until today they comprise about 30 percent of the SSI rolls—about 20 percent of these are eligible based on age and 11 percent on the basis of disability. Today about 80 percent of SSI recipients are eligible on the basis of disability; 56 percent of these are 18–64 years of age (SSA, 2001b).

The beneficiary population, especially in the SSI program, is diverse. Throughout the 1990s, the proportion of SSI awards each year for adults 18–64 who are noncitizens has ranged from 7 to 8 percent of the total (SSA, 2001b). The largest numbers come from Viet Nam, Mexico, and Cuba. Many of them have limited or no work experience and limited English proficiency (SSA, 2000).

The law provides uniform standards for citizenship and residency. However under certain circumstances, "qualified aliens" are eligible for SSI (some permanently and others for up to seven years). To qualify for SSI, someone who is not a U.S. citizen must be a qualified alien and meet one of certain additional requirements such as: a person lawfully admitted for permanent residence in the United States, a refugee, asylum seekers, or a person subjected to battery or extreme cruelty or whose child or parent has been subjected to such battery; or is a "qualified alien" who was lawfully residing in the United States and receiving SSI as of August

22, 1996, or who was living in the United States on August 22, 1996, and subsequently became blind or disabled (U.S. House of Representatives, 2000; SSA, 2001b). Legislative amendments in 2000 (P.L. 106-386) extended eligibility to noncitizens, regardless of their immigration status, as refugees if they are determined to be "victims" of "severe forms of trafficking in persons" (SSA, 2001b).

Age and Gender

The increases in applications and awards and a decrease in the number leaving the program have resulted in a dramatic growth in the number of beneficiaries on the rolls. This growth is due, at least in part, to an increase in the number of persons in the relatively younger ages entering the disability programs with fewer life-threatening impairments, resulting in increasing the duration of entitlement. As shown in Figure 2-3, the average age of persons awarded disability insurance benefits has been declining for both men and women, with a consequent increase in the duration of benefits. The average age of men awarded SSDI benefits declined from 54.5 in 1960 to 51.2 in 1980 and 49.6 in 2000, while the average age of women awarded SSDI benefits declined from 52.5 in 1960 to 51.1 in 1980 and 48.7 in 2000.

As seen in Table 2-4, in 1960 less than 1 percent of men and women who were awarded SSDI benefits were under 30 years of age, but by 2000, 6.8 percent of the men and 5.8 percent of the women were in this age range when awarded benefits. Similarly, the proportion of both men and women who were between 30 and 39 years of age when awarded benefits approximately doubled, while the proportion between ages 40 and 49 when awarded benefits also increased. In contrast, the proportion of men 50 to 64 years of age when awarded benefits decreased from about 75 percent in 1960 to nearly 57 percent in 2000; the proportion of women in this age group awarded benefits decreased from 70 percent in 1960 to almost 55 percent in 2000.

As increasing number of women have entered the labor force, the proportion of beneficiaries who are women has increased. Thus, in 1960, 78 percent of the 455,000 SSDI disabled worker beneficiaries were men and 22 percent were women, but by 2000, of the approximately 5 million SSDI disabled worker beneficiaries, the proportion who were men had declined to 56.6 percent and the proportion of women had increased to about 43 percent (SSA, 2001d).

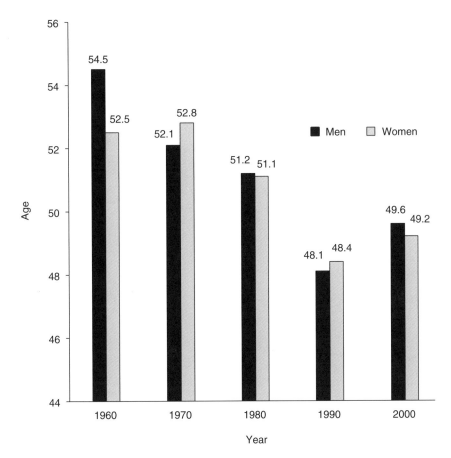

FIGURE 2-3 Average age of persons awarded SSDI benefits, by gender, 1960–2000.
SOURCE: SSA, 2001d.

Impairments

In the early years of the SSDI program, a much larger proportion of benefits were awarded on the basis of chronic diseases of aging. In recent years, as a result of legislative changes and court decisions, an increased number of awards have been based on impairments that occur much earlier in life such as mental disorders, including mental retardation. As shown in Figure 2-4, in 1981, 11 percent of SSDI disabled worker beneficiaries 18–64 years of age were awarded benefits on the basis of mental disorders including mental retardation compared with 24 percent in 2000,

TABLE 2-4 Percentage of Disabled Workers Awarded SSDI Benefits, by Gender, 1960–2000

Year	<30	30–39	40–49	50–64
		Age		
		Men		
1960	0.8	7.0	17.0	75.2
1970	6.7	7.6	16.6	69.1
1980	8.3	9.7	14.4	67.6
1990	10.9	16.9	18.9	53.3
2000	6.8	12.9	23.4	56.8
		Women		
1960	0.7	8.1	21.3	69.9
1970	4.2	6.3	17.1	72.4
1980	7.4	9.7	15.7	67.2
1990	8.5	16.3	22.9	52.3
2000	5.8	13.7	25.8	54.7

SOURCE: SSA, 2001d.

an increase of 118 percent. A similar distribution of impairments is noted for SSI working age beneficiaries, with 31 percent receiving benefits in 2000 because of mental disorders other than mental retardation and another 21 percent receiving benefits because of mental retardation.[4]

Between 1981 and 2000, the proportion of SSDI benefit awards based on circulatory conditions, the top ranked condition in the earlier year, declined by 52 percent. The proportion of persons awarded benefits on the basis of musculoskeletal conditions increased by 41 percent between 1981 and 2000. By 2000, musculoskeletal conditions had eclipsed circulatory conditions as the most common set of conditions associated with the award of SSDI benefits. One possible explanation is the aging of the baby boom generation cohorts (1946–1964) who are currently entering the ages of highest incidence of arthritis and back disorders (Helmick et al., 1995). In addition, rates of cardiovascular disease have declined over the past 10 to 15 years.

---

[4]1981 data for SSI comparable to those for SSDI are not available.

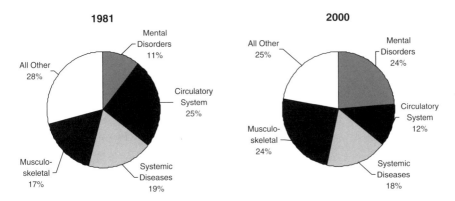

FIGURE 2-4 Percentage distribution of SSDI awards for adults aged 18–64, by diagnostic groups, 1981 and 2000.
SOURCE: SSA, 2001d.

Outreach Efforts

As mentioned in the previous section, mandated outreach activities to enroll persons in the SSI program contributed to growth in the programs in the early and mid-1970s, when a nationwide effort was launched to enroll eligible persons in the new SSI program that was enacted in 1972 and implemented in 1974. During the late 1980s, Congress again mandated a number of SSI outreach activities to facilitate applications by needy individuals with severe disabilities (U.S. House of Representatives, 2000). Beginning with earmarked appropriations in 1989, SSI outreach activities became a priority for SSA. In addition to its own effort, SSA promoted outreach through a series of grants to the private sector (Muller and Wheeler, 1995). Some who applied for SSI were found to have enough covered work experience to qualify for disability insurance benefits concurrently with SSI.

Cost Shifting by States

In times of poor economy, cuts made in state and locally funded general assistance and other welfare programs result in shifting the burden from state and local programs to federal programs. Welfare agencies routinely refer persons to SSA's disability programs. During 1989–1992, such cost shifting may have contributed to the acceleration of applications and awards particularly in the SSI programs (Rupp and Stapleton, 1995). Deinstitutionalization of persons with mental disorders and mental retardation and other disabilities who were previously cared for in and financed by state hospital systems also led to an increase in the SSI claims.

Legislative and Regulatory Changes and Court Decisions

As in the early years of the programs, legislative and regulatory changes and court decisions continue to play a crucial role in extending or restricting the scope of the disability programs. For example, Congress and the courts required revision of the medical and functional criteria and new evidentiary requirements used in determining eligibility for disability benefits. In 1984, Congress required SSA to review and update the Listings of Impairments and related regulations resulting in significant increases in awards of benefits for mental impairments in the late 1980s. The 1996 legislation dropped drug abuse and alcoholism as a contributing factor material to finding disability. The Welfare Reforms and legislative changes with respect to qualifying noncitizens for SSI benefits also led to changes in applications for benefits. Court decisions have had a major impact on the programs by modifying and liberalizing the adjudication standards. The publicity created by court cases increased public awareness and perception of easier standards, which in turn may have led to growth in applications for benefits.

Structural Changes in the Labor Market

Structural shifts in the economy have an uncertain impact on workers with disabilities and can influence the types of impairments that result in work disability. For instance, the shift from manufacturing to service industries and occupations and the emergence of new terms of employment have been emerging over the past several decades. The changing demands of work also limit employment prospects for individuals whose social and adaptive functioning is impaired by mental disorders. The current labor market places emphasis on cognitive and technical skills, advanced education, and the ability to communicate and interact with others. People with disabilities, especially those with mental impairments, have poor employment prospects in such a market.

Rupp and Stapleton (1995) argue that business cycle effects overwhelm the effects of economic restructuring. Their study found a negative effect of restructuring for the SSDI only, but it was small and not replicated for other program categories; they found no significant long-term effect of economic restructuring. Although the short-term effect of economic restructuring may increase applications for benefits, the long-term effect may be to decrease applications because workers in service sector jobs are less susceptible to disabling injuries, at least as far as physical impairments are concerned.

Changes in Economic Conditions

As in the 1970s, the economic downturn of 1990–1991 contributed to the rapid increase in applications and awards in the disability programs (Stapleton et al., 1998). In times of poor economy and high unemployment, low-wage workers with disabilities are more adversely affected than others in the workforce (Yelin, 1992). Lack of training, unavailability of adequate alternative employment, and prospects of losing medical coverage lead to increased applications for disability benefits. In the 1980s, despite poor economic conditions at the time, applications and awards declined as a result of programmatic changes in the SSDI and SSI programs.

Labor Market Dynamics[5]

In 1960, almost all working age (18–64) men were in the labor force, while only a minority of women of these ages were in the labor force. Since then, labor force participation rates among men fell, dramatically so among men 55–64 years of age, the decade prior to entitlement to full Social Security retirement benefits. At the same time, labor force participation rates among women have increased. In 1960, 66.8 percent of all working age persons were in the labor force. Primarily because of the substantial increases in employment among women, the overall labor force participation rate among working age persons increased to 79.0 percent as of 1998, or by more than 18 percent in relative terms. In 1960, 93.2 percent of all working age men were in the labor force. However, male labor force participation rates fell dramatically, particularly after 1970, before stabilizing in the last few years. By 1998, only 86.3 percent of all working age men reported being in the labor force, a decline of 7 percent in relative terms. The employment picture among women is the reverse of that observed among men, with substantial increases in labor force participation rates among women. Thus, between 1960 and 1998, labor force participation rates among all working age women rose from 42.7 to 72.0 percent, or by 69 percent (calculations using data from BLS, 1999, Jacobs, 1999).

Employment patterns among persons with disabilities reflect these overall trends among men and women (Table 2-5).[6] Among all working

---

[5]Much of the information in this section is drawn from the background paper commissioned from Edward Yelin and Laura Trupin for use by this committee. The committee appreciates their contribution. The full text of the paper can be found in Part II of this report

[6]Throughout this section the National Health Interview Survey definition of disability is used, that is, those persons who report a limitation in the ability to do usual major activity, in the kind or amount of activity, or in outside activities.

TABLE 2-5 Labor Force Participation Rates (percent) of Persons with and Without Disabilities, by Gender, United States, 1983–1999

Gender and Disability Status	Year		Percent Change 1983–1999
	1983	1999	
All persons	75.0	78.6	4.8
With disabilities	48.6	41.5	−14.6
Without disabilities	79.1	82.8	4.7
All men	87.2	85.9	−1.5
With disabilities	60.0	44.9	−25.2
Without disabilities	91.5	90.4	−1.2
All women	63.8	71.6	12.2
With disabilities	38.0	38.5	1.3
Without disabilities	67.6	75.5	11.7

SOURCE: Adapted from Trupin et al., 1997, and reprinted with permission from Yelin, 1999.

age persons with disabilities, labor force participation rates decreased from 48.6 to 41.5 percent between 1983 and 1999, a decline of 14.6 percent. This overall decrease is the net effect of a huge decrease among men with disabilities—from 60.0 percent in 1983 to 44.9 percent in 1999, or by 25.2 percent—and a slight increase among women with disabilities from 38.0 percent in 1983 to 38.5 percent in 1999. Compared to men without disabilities, men with disabilities experienced a larger relative decrease in labor force participation rates (25.2 versus 1.2 percent, respectively). Compared to women without disabilities, women with disabilities experienced a much smaller relative increase in these rates (1.3 versus 11.7 percent, respectively). This is consistent with the hypothesis that persons with disabilities, like those facing discrimination on the basis of age, gender, or race, are prone to a last hired, first fired phenomenon.

Labor Market for Persons with Disabilities

Although among all working age persons, including men (and even extending to men 55 to 64 years of age), labor force participation rates have increased in the last few years, two recent studies indicate that persons with disabilities have not shared in this increase (Bound and Waidmann, 2000; Burkhauser et al., 2000), while another study finds no change (Levine, 2000). Thus, although labor force participation rates among persons with disabilities reflected the trends affecting all workers over the past two decades, there is now at least equivocal evidence that, despite the passage of the Americans with Disabilities Act of 1990 (ADA), the employment picture among persons with disabilities diverged from that among the remainder of the working age population at the end of the 1990s. Studies conducted in recent years have suggested that the ADA may have unintentionally harmed rather than helped workers with disabilities (DeLeire, 2000a, 2000b; Acemoglu and Angrist, 1998). The ADA was enacted to remove barriers to employment for people with disabilities by banning discrimination and requiring employers to provide accommodations. However, the costs of complying with the Act and fear of litigation may reduce the demand for their labor and undo its intended effect. Bound and Waidmann (2001) using the Current Population Survey (CPS) data from 1989 to 1999 found little evidence indicating much of a role for the ADA, but argue that increases in Social Security disability benefits played an important causal role in the growth of the population on disability rolls and can account for the decline in employment of working age men with disabilities during the period. Others also have indicated that the increasing program generosity and worsening labor market conditions increase the option value of disability applications (Autor and Duggan 2001).

The literature on employment among persons with disabilities suggests that their labor force participation rates appear to reflect more general labor market dynamics (Yelin, 1992, 1999; Stapleton et al., 1998). Consistent with this observation, during the slack labor markets in the 1970s and late 1980s, there were substantial increases in the number of people applying for Social Security Disability Insurance benefits.

CONCLUSION

The impact of any one factor on the demand for and provision of disability benefits is difficult to determine. In addition to the factors already discussed above, other factors also may have led to the growth of the disability programs at different times. These include record low termination rates of beneficiaries, public perceptions about the ease of qualifying for benefits, and access to medical care and its role in influencing choice between work and acceptance of disability benefits.

As stated earlier in the chapter, the disability rolls are projected to grow over the coming decades as the baby boom generation reaches the ages of increased likelihood of developing disabilities. This increase in workloads will make it increasingly important for SSA to have clear and workable policies, rules, and guidelines to operate its programs and to project future growth. The gradual increase in full retirement age from 65 to 67 years also means that disabled workers may remain on the rolls for two additional years before converting to Social Security retirement. An improved understanding of the dynamics of the programs and the factors that influence them is required. At this time, little is known with certainty about what contributes to disability trends and to what degree. Ongoing and future research using new data sources, such as the data that will be generated by the National Study of Health and Activity and other SSA administrative files, should provide relevant information about disability program participation and cost and other related policy issues.

Moreover, as aptly stated by Burkhauser et al. (2001), "no studies have been able to satisfactorily disentangle the impact of demand side factors related to the passage of the ADA or changes in the mix of jobs in the economy in the 1990s from supply side factors related to changes in the ease of access to SSDI and SSI benefits or to a deduction in the share of jobs that provide private health insurance, which would discourage work among the population with disabilities." Research is needed to delineate the magnitude of the various effects in order to understand the causes of recent declines in employment among people with disabilities. Only then can policies be developed to reverse the trend.

3

Conceptual Issues in Defining Work Disability[1]

This chapter discusses the meaning of the term "disability" and the relationship between the generic concept of disability and the term "work disability." The meaning assigned to these terms depends on the uses to be made of the concepts. As indicated in Chapter 1, the primary concern of the present study is with disability as used in the Social Security Administration (SSA) disability programs—the inability to engage in any substantial gainful activity because of physical or mental impairments that are medically determinable. However, in order to place the concept of work disability in perspective, first, definitions of disability are examined in a wider context.

CONCEPTS OF DISABILITY AND WORK DISABILITY

Defining disability has occupied the attention of many individuals and groups in the United States for many years. The problem of defining disability to determine eligibility for income maintenance, the perspective of the SSA, may be viewed in the context of the researchers and scholars who have considered this question in different cultures and in different

---

[1]Much of the information in this chapter is drawn from the background paper "Conceptual Issues in the Measurement of Work Disability," commissioned by the committee from Alan Jette and Elizabeth Badley for its Workshop on Survey Measurement of Work Disability held in 1999. The committee appreciates their contribution. The full text of the paper can be found in the workshop report (IOM, 2000) and is reproduced in Part II of this report.

contexts. Considerable conceptual controversy exists, growing out of different traditions that have generated several major disability frameworks found in contemporary literature. Processes of social change—including, for example, civil rights movements and development of assistive technology—have contributed to the emergence of varied, even competing, frameworks. Nonetheless, across the several different schools of thought, one can identify scholarly consensus on what constitutes the phenomenon of disability, which is the first step to achieving a common language.

How a society defines and treats persons with a limitation in ability or disability has roots in many different cultures. Contemporary Western thought can be traced to functionalism as expressed in Talcott Parsons' definition of illness as "a state of disturbance in the normal functioning of the total human individual including both the state of the organism as a biological system, and of his personal and social adjustments" (Parsons, 1951, p. 431). This sociological perspective was the basis for definitions of disability focusing on functional status for many decades, resulting in an emphasis on the individual and individual adjustment with less attention to the extrinsic or environmental factors that influence a person's ability to work or engage in meaningful activity. The Americans with Disabilities Act (ADA), for example, defines the term disability "with respect to an individual—(A) a physical or mental impairment that substantially limits one or more of the major life activities of such individual; (B) a record of such an impairment; or (C) being regarded as having such an impairment" (ADA, 2001).

Researchers have attempted to define disability by designing models that document the process of becoming disabled. Some of these models include: the Functional Limitation Paradigm (Nagi, 1965, 1979); the International Classification of Impairments, Disabilities, and Handicaps (ICIDH) (WHO, 1980), recently revised and renamed the International Classification of Functioning, Disability and Health (ICF); the Institute of Medicine (IOM) models (IOM, 1991, 1997b); and variations from other authors in many different contexts (Patrick and Peach, 1989; Verbrugge and Jette, 1994). While each of these models suggests a theoretical definition of disability, none offers a detailed operational definition, although the extensive classification system of the ICF permits multiple coding for individuals.

Scholarly definitions differ among themselves, but they share one thing in common. As long as they are confined to the abstract, theoretical level, they are free to add to, subtract from, or modify any of their terms and conditions, subject only to canons of logic and the scholar's considered judgment. However, once the definitions are applied under real-world conditions, they necessarily operate under constraints of one sort or another, leading to numerous definitions used in public laws and private contracts offering different kinds of benefits or services or in a survey context to measure inability to undertake major activities of daily living. No single definition is feasible or desirable that will fit all purposes of assessment.

Consider the main purposes to which definitions of disability are applied. A major purpose of most relevance to this report is eligibility for cash benefit programs such as Social Security Disability Insurance (SSDI) and Supplemental Security Income (SSI). Under these programs, the definition is used as a screening device. People who meet the definition receive the benefit, while those who fail to meet the definition are denied access to the benefit. The immediate and obvious constraint is that the screening of candidates for access to the benefit involves costs in terms of time of both applicants and screeners. The nature and type of these constraints under which the administering agency chooses to operate will depend on the value of the benefit that is being offered and the number of applicants. For example, the situation is obviously quite different comparing the benefits offered to applicants for a handicapped parking program and for the Social Security disability programs. No doubt the handicapped parking space is valuable to the applicant, but its value surely pales in comparison to that of income maintenance that may last a lifetime.

The monetary value of the benefit is relevant, but the resources available to screen the applicants are also important. In the Social Security disability programs, the benefits are quite valuable, whereas the resources devoted to screening applicants are limited in relationship to the demand for benefits. As the statutory definition makes clear, a person is considered "work disabled" based on the existence of a medical impairment or injury that precludes substantial gainful activity (SGA). With millions of applicants each year, SSA has to resort to administrative shortcuts in making decisions. Consequently SSA uses Listings of Impairments (Listings) as a critical early decision step to award or deny benefits. These Listings consist of medical evidence of more than 100 conditions that are considered to be of such severity that the condition can be presumed to constitute work disability regardless of the individual's age, education, previous occupation, or other relevant factors.

Similar problems can be seen in the administration of other benefit programs such as workers' compensation. In that state program, various states use different methods to judge eligibility for benefits. One benefit found in most of the state jurisdictions is for a permanent partial disability. An example of this is the American Medical Association Guides to the Evaluation of Permanent Impairment (AMA, 1993), which is a standardized system for translating the extent of an injury of a body part into a percentage of disability of the whole person. This type of system has been used for the assessment of compensation payments, including workers' compensation. Such a benefit is paid to a worker who is left with a partial disability after an accident or exposure to an occupational illness. Some states make these awards on the basis of a demonstrated actual wage loss that occurs because of the effects of the injury or as a sequel to the injury or exposure.

Other jurisdictions make these awards on the basis of the identifiable medical impairment or loss of limb, while still others make their decisions on the basis of what they term "loss of wage-earning capacity." Essentially the latter concept uses the evaluation of the impairments and then modifies them according to the age and occupation of the worker. In spite of these differences, the theoretical basis for the awards is the same in all jurisdictions (Berkowitz and Burton, 1987). The awards are made because of the economic losses suffered by the worker by reason of the partial inability to participate in the labor force. The methods of evaluating these losses differ as each state seeks some method of approximating the concept in an administratively feasible manner in these large benefit programs. SSA faces the same necessity to find some easily understood method of making decisions in its disability programs.

In addition to benefit programs, the other main purpose of disability definitions that is most relevant to this report is in the field of surveys that have their own space and time constraints. The broad conceptual definitions are necessarily shortened, and particular portions emphasized, as survey personnel seek to fit their questions into the prescribed few lines or few minutes of time.

Recognizing these real-world constraints does not diminish the importance of the theoretical abstract concepts on which the survey or benefit definitions are based. It is the theory that gives us the objectives to which the program strives. Whether disability is a purely medical concept that can be decided by examining an individual or whether it is a relational concept is an issue that first must be explored on a theoretical level. The process of applying these concepts for providing benefits or conducting a survey may obscure the theoretical foundations, but they are surely present.

MAJOR SCHOOLS OF THOUGHT

Several schools of thought have defined disability and related concepts. Jette and Badley's paper "Conceptual Issues in the Measurement of Work Disability" (IOM, 2000) examines in detail the different concepts or definitions including those set forth by Nagi, the World Health Organization (WHO), the IOM, Verbrugge and Jette, and others. This chapter highlights some of the key points made in that paper. As discussed by Jette and Badley, the major differences in these frameworks lie more in the terms used to describe disability and related concepts, and in the placing of the boundaries between these concepts, than in their fundamental content. Both Nagi's Disablement Model and WHO's ICIDH frameworks have in common the view that overall disablement represents a series of related concepts that describe the consequences or impact of a health condition, in an interaction with a person's environment, on the person, on the person's activities, and on the wider participation of that person in society. After reviewing terms within each framework, the two major models along with their major derivatives can be compared and contrasted and their relationships more generally to the concept of work disability can be explored.

According to the conceptual framework of disability developed by sociologist Saad Nagi (1965), disability is the expression of a physical or a mental limitation *in a social context*. In striking contrast to the SSA's definition of work disability as inability to work as a consequence of a physical or mental impairment, Nagi specifically views the concept of disability as representing the gap between a person's capabilities and the demands created by the social and physical environment (Nagi, 1965, 1976, 1991). This is a fundamental distinction of critical importance to scholarly discussion and research related to disability phenomena.

According to Nagi's own words:

> Disability is a limitation in performing socially defined roles and tasks expected of an individual within a sociocultural and physical environment. These roles and tasks are organized in spheres of life activities such as those of the family or other interpersonal relations; work, employment, and other economic pursuits; and education, recreation, and self-care. Not all impairments or functional limitations precipitate disability, and similar patterns of disability may result from different types of impairments and limitations in function. Furthermore, identical types of impairments and similar functional limitations may result in different patterns of disability. Several other factors contribute to shaping the dimensions and severity of disability. These include (a) the individual's definition of the situation and reactions, which at times compound the limitations; (b) the definition of the situation by others, and their reactions and expectations—especially those who are significant in the lives of the person with the disabling condition (e.g., family members, friends and associates, employers and co-workers, and organizations and professions that provide services and benefits); and (c) characteristics of the environment and the degree to which it is free from, or encumbered with, physical and sociocultural barriers. (Nagi, 1991, p. 315)

Nagi's definition stipulates that a disability may or may not result from the interaction of an individual's physical or mental limitations with the social and physical factors in the individual's environment. Consistent with Nagi's concept of disability, an individual's physical and mental limitations would not invariably lead to work disability. Not all physical or mental conditions would precipitate a work disability, and similar patterns of work disability may result from different types of health conditions. Furthermore, identical physical and mental limitations may result in different patterns of work disability.

Nagi's Disablement Model has its origins in Functionalism identified most closely with Talcott Parsons (1951). In the early 1960s, as part of a study of decision making in the SSDI program, Nagi (1964) constructed a framework that differentiated from three other distinct yet interrelated concepts: active pathology, impairment, and functional limitation. This conceptual framework has come to be referred to as Nagi's Disablement Model.

In their work on the disablement process, Verbrugge and Jette (1994) maintained the basic Nagi concepts and his original definitions. Within the dimension of disability however, they categorized subdimensions of social roles that can be considered under Nagi's concept of disability. Some of the most commonly applied dimensions include activities of daily living, instrumental activities of daily living, paid and unpaid role activities, social activities, and leisure activities. Within their framework, work disability is clearly delineated as a specific subdimension under the concept of disability.

A further elaboration of Nagi's conceptual view of the term disability is contained in *Disability in America* (IOM, 1991) and in a more recent IOM revision of the disablement model highlighted in a report titled *Enabling America: Assessing The Role of Rehabilitation Science and Engineering* (IOM, 1997b). The 1991 IOM report uses the original main disablement pathways put forth by Nagi with minor modification of his original definitions. That report makes two important additions to the Disablement Model: the concepts of "secondary conditions" and "quality of life." In 1997, in an effort to emphasize that disability is not inherent in the individual (as defined by SSA) but rather is a product of the interaction of the individual with the environment, the IOM issued the second report, *Enabling America*, where it referred to disablement as "the enabling–disabling process." This effort was an explicit attempt to acknowledge within the disablement framework itself that disabling conditions not only develop and progress but can be reversed through the application of rehabilitation and other forms of explicit intervention.

The ICIDH, now revised and renamed the International Classification of Functioning, Disability and Health (WHO, 2001), has moved away from a "consequence of disease" classification (WHO, 1980) to a "components of health" classification that includes a list of environmental factors that describe the context in which individuals live. Components of functioning and disability include the body component, the activities and participation component, and contextual factors, including a list of environmental factors and personal factors. A person's functioning and disability are conceived as dynamic interaction between health conditions and contextual factors. The basic construct of environmental factors is the facilitating or hindering impact of features of the physical, social, and attitudinal world.

Concept of Social Roles

Social roles, such as being a parent, a construction worker, or a university professor, are basically organized according to how individuals participate in a social system. According to Parsons, ". . . role is the organized system of participation of an individual in a social system" (Parsons, 1958, p. 316). Tasks are specific activities through which the individual carries out his or her social roles. Social roles are made up of many different tasks, which may be modifiable and interchangeable. Some social roles are more flexible than others; that is, there is greater societal acceptance for modifying and interchanging tasks in some roles than others.

Fundamental to differentiating the concept of disability from those of pathology, impairment, and functional limitation is consideration of the difference between concepts of attributes or properties on the one hand and relational concepts on the other (Cohen, 1957).

To take the example of limitation in the performance of one's work role, work disability typically begins with the onset of one or more health conditions that may limit the individual's performance in specific tasks through which an individual would typically perform his or her job. The onset of a specific health condition—for example, a stroke or a back injury—may or may not lead to actual limitation in performing the work role—a work disability. The development of work disability will depend, in part, on the extent to which the health condition limits the individual's ability to perform specific tasks that are part of one's occupation, and alternatively, the degree of work disability may depend on external factors, for example, workplace attitudes, such as flexible working hours, that may restrict employment opportunities for persons with specific health-related limitations. Work disability also might be affected by accessible modes of transportation to the workplace, environmental barriers in the workplace, or willingness to modify the individual workstation to accommodate a health condition. Viewed from the perspective of role performance, the degree of work disability could be reduced by improving the individual's capacity to accomplish functional activities (a very traditional view of rehabilitation) or by manipulating the physical or social environment in which work occurs.

The fundamental conceptual issue of concern is that a health-related restriction in work participation may not be solely or even primarily related to the health condition itself or its severity. In other words, although the presence of a health condition is a prerequisite, "work disability" may be caused by factors external to the health condition's impact on the structure and functioning of a person's body or the person's accomplishment of a range of activities.

The Dynamic Nature of Disability

The earliest disablement models represented by Nagi (1965) and the ICIDH-1 formulation (WHO, 1980) presented the disablement process as more or less a simple linear progression of response to illness or consequence of disease. One consequence of this traditional view is that disabling conditions have been viewed as static entities (Marge, 1988). This traditional, early view of disablement failed to recognize that disablement is more often a dynamic process that can fluctuate in breadth and severity across the life course. It is anything but static or unidirectional.

More recent disablement formulations or elaborations of earlier models have explicitly acknowledged that the disablement process is far more complex (IOM, 1991, 1997b; Verbrugge and Jette, 1994; Badley, 1995; WHO, 1997, 2001; Fougeyrollas, 1998). These more recent studies note that a given disablement process may lead to further downward spiraling consequences. IOM (1991) uses the term *secondary conditions* to describe any type of secondary consequence of a primary disabling condition. IOM (1991) also included quality of life in the conceptual model, although little attention was given to how to define this concept or make it operational for persons with disabilities. Patrick (1997), in rethinking preventive interventions for people with disabilities, focused on opportunity as the intersection between the total environment and the disabling process and defined opportunity as the four goals of the ADA, including economic self-sufficiency and full participation, which are highly related to work. Quality of life is viewed as people's perceptions of their position in life in the context of culture and personal goals and expectations. Quality of life is the final outcome and is influenced by all aspects of the total environment, experience with health care, the disabling process, and level of opportunity.

The Concept of Work Environment

The social context for disability assessment concepts is implied in most schools of thought. The social context for SSA is the work environment. Establishing whether a person is capable of performing past relevant work or any type of substantial gainful activity in the national economy is part of the disability decision process. SSA has been using the Department of Labor's (DOL's) Dictionary of Occupational Titles (DOT) and selected characteristics of occupations as a basis for defining the work environment. SSA plans to replace the DOT as a description of work environment with the Occupational Information Network (O*NET) database of work requirements that is being developed by the DOL. (DOT and O*NET are discussed further in Chapter 6.) The importance of these descriptions is the interaction between the concept of an individual's impairment and the requirements of work as influencing the ability to work in the national economy.

Relating Definitional Issues to the Measurement of Work Disability

The underlying structure of models of disablement, as discussed above, maps a pathway between a health condition or injury and the ensuing work disability. Close inspection of the definitions given above suggests that a number of steps can be identified in the pathway between the health condition and the social consequences described as work disability. At a micro level there are pathological changes in the body and impairment in the structure and functioning of organs and body systems. There may be an impact on the activity of the person, ranging from simple movements, to basic activities of daily living, to instrumental activities of daily living, and so on. These then can contribute to the individual's capacity to perform more complex social roles, and ultimately the person's participation in all aspects of society can be adversely affected. Work is one such social role.

As indicated earlier, work disability is a function of whether the person can perform specific work-related tasks and of external factors. From the point of view of the measurement of work disability, it may be useful to distinguish between the degree of difficulty a person may have in carrying out an activity and the other factors (such as barriers in the environment, attitudes of employers or coworkers, and other restrictions) that might prevent the performance of that activity in daily life. In this way, the levels of impact described within the conceptual models are of importance because they allow us to locate where many of the current types of assessment of work disability might fit in.

Discrete or Continuous Phenomena

Disability is commonly presented as an all or nothing phenomenon, either a person "is disabled or not." In reality, disability in particular roles or activities is usually encountered in terms of degree of difficulty, limitation, or dependence, ranging from slight to severe. The question then becomes: Where on the disability spectrum is the threshold that determines whether a person has a disability or work disability? The question needs to take into account any assistive devices or accommodations that the person may have. In the current context, work participation is often determined as being an end point, in that people either have a work disability or they do not. In reality, the situation is likely to be more complex. For example, many people with functional and activity limitations may continue to work, but their labor force participation may be compromised in some way by the condition, including the opportunity to work. To the extent that it is, these people might be said to have some degree of work disability. In measuring work disability, a clear definition of the threshold used needs to be made.

Duration or Chronicity

There is a pervasive assumption that work disability is a long-term state. Stereotypes about disability are dominated by the archetype of a person who uses a wheelchair. Embedded in this is the notion of some disabling event, a period of adjustment and rehabilitation, and then the resumption of as full a life as possible with the assistance of any necessary assistive devices or accommodations. With much impairment, the reality of disability is somewhat different. The majority of individuals in the working age population with long-term activity restriction report that this restriction is due to musculoskeletal, circulatory, or respiratory disorders (LaPlante et al., 1996). These conditions may also be associated with varying degrees of "illness," so that it is not just an issue of physical performance. Other considerations are pain, fatigue, and other symptoms. Many of these conditions are episodic in nature and may have trajectories of either deterioration or recovery (the latter being less common). Apart from any environment barriers or facilitators, the day-to-day or month-to-month experience of disability may be variable and may need to be taken into account in any measurement scheme.

In summary, researchers have attempted to define disability by designing models (or paradigms) that document the process of becoming disabled. While each of these models suggests a theoretical definition of disability, none offers a detailed operational definition. All definitions agree, however, in viewing disability as an intersection between the individual intrinsic situation and the external environment that places demands and provides opportunities for individuals with disabilities.

CONCLUSION

Due to the necessity for finding some economical administrative methods of deciding eligibility in this mass production benefit program, in the majority of applications an adult is considered work disabled based solely on the existence of a medical impairment or injury that is presumed to preclude SGA. The foundation of the current work disability determination process, however, rests on medical evidence of more than 100 medical conditions (organized into Listings of Medical Impairments) that are considered to be of such severity that the condition can be presumed to constitute work disability. The determination process generally does not take into explicit account the relation of the individual within the work context.

The problem with this approach with regard to the definition and determination of work disability, as indicated by the above discussion of disability concepts and frameworks, is that a one-to-one relationship is unlikely between the presence of medical conditions and the resultant impairments and subsequent disability in substantial gainful employment. The presumption within the current SSA determination process that work disability is a direct reflection of the severity of the person's medical condition and/or resultant impairment may have outlived its usefulness. In light of the ADA, medical advances, and new developments in technology, more attention needs to be paid to the environment, particularly in the context of work disability and vocational rehabilitation.

The committee recognizes the administrative difficulties that might be involved and that such attention may require drastic shifts in the orientation of the Social Security disability programs. Primary attention may have to shift to ways to influence the environment in which the applicant might work and to "return to work" activities. In the face of these challenges in incorporating contemporary concepts of disablement that include the dynamic nature of work, functioning, and health, SSA should undertake research focused on the relationship between the individual and the work environment and the evaluation of vocational factors as they affect work disability.

Recommendation 3-1: The committee recommends that the Social Security Administration develop systematic approaches to incorporate economic, social, and physical environmental factors in the disability determination process by conducting research on

- the dynamic nature of disability;
- the relationship between the physical environment and social environment and work disability; and
- understanding the external factors affecting the development of work disability.

If such research is fruitful, incorporating such changes in the Social Security disability determination process will begin to move it away from a predominantly medically driven approach to consider factors beyond physical, sensory, cognitive, or emotional impairments and may ultimately involve changes in SSA's implementing regulations.

As this chapter has shown, a full understanding of work disability needs to take into account the individual's circumstances and the social and physical environment of the workplace. *The research challenge is to apply the insights provided by the current models of disability to come to a common understanding of work disability concepts, and to understand the dynamics of the pathway between health conditions and work disability. Researchers need to find ways to incorporate an understanding of external factors influencing the development of work disability into future measurement strategies.*

4

Survey Measurement of Disability

This chapter first provides a brief summary of the general features of the National Study of Health and Activity (NSHA) as planned and the experience to date in the planning and development of the survey. The remainder of the chapter is structured around the key statistical issues of measurement facing such complex surveys. For each issue the chapter describes the basis of the issue, gives examples of the issue as illustrated by the NSHA, and then draws more general implications for the Social Security Administration (SSA) research agenda in work disability.

As stated in Chapter 1, NSHA is a response to the recommendation made by the Board of Trustees of the Federal Old Age and Survivors Insurance and Disability Insurance Trust Funds (DHHS, 1992). SSA considers NSHA as the cornerstone of its long-term disability research plan aimed at understanding the growth of the disability programs. It is also needed to answer policy and research questions about the nature and extent of disability in the United States. SSA also needs to know the magnitude and characteristics of the population with disabilities who may be eligible for benefits and the factors that keep them employed. It needs answers to these and other questions in order to project future trends in its disability programs with a degree of confidence.

A major component of the committee's deliberations has been to evaluate the NSHA—its information goals, the process of developing measurements to meet these goals, the method of data collection, and the sample selection and allocation required to adequately represent the potential recipients of disability benefits. The committee has focused on matters of measurement issues to meet the Social Security Administration's information needs, on the adequacy of the research design, and the implementation plan for the NSHA. The committee issued two interim reports (IOM, 1997a, 1999b) on its findings and conclusions based on its review. The first interim report provided a preliminary review of the general features of the proposed survey design, data collection plans, coverage, sampling plans, and operational decisions as described in the scope of work prepared by SSA in the draft request for proposals (RFP) for the conduct of NSHA. The committee believed that SSA needed to make important decisions about the survey design, the research and development work for the survey, and other basic features before issuing an RFP for the survey. It also discussed some of the limitations as they related to the efficiency of the sampling plan in terms of accepted statistical principles and practices. The committee's third interim report reviewed and provided guidance on the sample design, instruments and procedures, and response rate goals for the pilot study. It also commented on the time line established by SSA for initiation of each phase of the survey. Both reports provided SSA with specific and detailed guidance on various aspects of the survey. SSA has responded by altering various features of the survey. All of the committee's recommendations made in these reports can be found in Appendix C.

THE NATIONAL STUDY OF HEALTH AND ACTIVITY

The National Study of Health and Activity is a complex, national sample survey designed to estimate the number and characteristics of a broad range of people with disabilities that affect their ability to work and carry out activities of daily living. SSA has contracted with Westat to conduct the survey. As originally conceived, the principal information goals of the NSHA were to

1. Estimate the total number and characteristics of people who are severely enough impaired that, but for work or other reasons,[1] they would meet SSA's statutory definition of disability. (This group would represent the universe of potentially eligible non-beneficiaries who could apply and meet the current criteria, but who are not now receiving benefits.)

---

[1]The term *work* for SSA's purposes refers to substantial gainful employment, which is generally about $780 per month for 2002. Other reasons for not receiving benefits include people who have chosen not to apply for disability, who have too many assets, who rely on family for support, or who are unaware of the program.

2. Identify the number and characteristics of people who are not eligible under the current SSA definition of disability, but who could be included as a result of any changes in the disability decision process.
3. Identify the factors (e.g., accommodations, social support, and other factors) that permit persons with similar impairments, who could qualify for benefits, to continue working.
4. Examine the variables needed to monitor and assess in a cost-effective manner future changes in the prevalence of disability.

In addition, SSA plans included simulating the disability applicants' folders developed at the Disability Determination Service (DDS) level using measures collected from the survey.

While the NSHA was being developed, efforts to redesign the disability decision process were on a parallel but separate track. NSHA assumed an additional role of evaluating the proposed redesigned process and of serving as a source for testing functional assessment instruments and the decision process itself. The original goals and design of the study were modified to accommodate an additional role for the NSHA. This part of NSHA design was subsequently dropped when SSA made the decision to no longer pursue the redesign initiative.[2]

More recently the survey has assumed an additional role to obtain data to explore if people with disabilities support SSA's Disability Employment Strategy, an initiative designed to encourage people with disabilities to continue to work or to leave the rolls and return to work by providing incentives to keep more earned income relative to benefits. SSA is currently assessing the impact of the Ticket to Work program that would allow Social Security Disability Insurance and Supplemental Security Income beneficiaries to keep $1 for every $2 earned. Another information goal added for the survey is to identify the effects of planned or possible increases in the retirement age on the disability program.

General Features of the NSHA Design

Sample Design

The sample design for the NSHA is driven by the following four core objectives (Westat, 1999b, p. 5). The design should yield samples of sufficient size to produce statistically precise estimates for

---

[2]In late 1999, SSA decided to abandon the redesign initiative. Chapter 6 of this report discusses the redesign initiative and the decision by SSA to shift away from it.

1. various subgroups of working age people with severe enough disabilities to be eligible for disability benefits for SSA purposes if they applied;
2. "borderline" group of people, with disabilities sufficient to permit estimates of the number and characteristics of those who might become eligible, or cease to be eligible, if the current SSA disability decision criteria are altered;
3. people with only mild or no disabilities, sufficient to permit comparisons with the population with disabilities on measures of physical and functional performance and medical conditions in the population; and
4. people currently receiving disability benefits under the Social Security Disability Insurance (SSDI) program and/or the Supplemental Security Income (SSI) program.

The sample for the NSHA is a dual-frame, multistage, stratified probability sample design. The first stage is a stratified sample of primary sampling units (PSUs) selected with probability proportional to size. Within the PSUs, households with persons 18–69 years of age are subsampled at rates designed to yield a nationally representative sample.

Sample Sizes

The sample sizes appear to be driven primarily by the first objective and by cost considerations. With those two factors in mind, SSA set a target to identify a sample of about 3,090 nonbeneficiaries with severe disabilities (the likely eligible group) out of a total sample of about 5,665 persons. Severe impairments are relatively rare in the general population. In fact, the severity and prevalence of a disabling condition are inversely related; the higher the prevalence of a condition, the lower the severity, and vice versa (LaPlante, 1991). Because SSA's eligibility criteria tend to filter out people with less severe disabilities, SSA is faced with many low-prevalence disabling conditions, all of which cannot be screened adequately into the sample. The exceptions may be mental conditions and low-back conditions. SSA is cognizant of this situation; therefore it has built into its sampling plan provision for oversampling persons with severe disabilities.

Accordingly, the sample was conceived to contain

- a "core" group of nonbeneficiaries with severe disabilities (about 3,090);
- persons with significant but lesser disabilities, the "borderline" cases (about 1,545);

- nondisabled persons (about 515); and
- current SSDI and/or SSI disability beneficiaries, who will be included primarily for the purpose of benchmarking the distinctive characteristics of the core group (about 515).

The first group, a core group of nonbeneficiaries, would consist of persons whose impairments are severe enough that they would likely be eligible for disability benefits if they applied. Other subgroups—current beneficiaries, people with lesser impairments (the "borderline" group), and nondisabled—are to be included in the survey to ensure full coverage as well as to provide the data needed to meet the NSHA objectives.

Data Collection Plans

Data collection for the NSHA involves

- a screening interview of a household respondent;
- a personal interview and physical performance tests;
- an extensive medical, and if needed, psychological examination; and
- a series of core and special medical tests.

In addition, SSA would obtain all medical evidence of record identified by the respondent and by third party reports on all persons in the sample to supplement information from the interviews and medical examinations in order to determine if the person meets SSA's current definition of disability.

Response Rates

SSA's assumptions about the sample size that would have to be screened in order to obtain the required 5,665 persons distributed disproportionately in the four strata for the various components were based on achieving the following response rates:

- 90 percent for the initial screening interview;
- 90 percent for the subsequent in-person interview and medical examination among those screened; and
- 80 percent overall response rate for the combined interview and medical examination components.

Assuming that these high response rates could be achieved, Westat estimated that a sample of about 98,095 persons in about 57,712 households would be sufficient to yield 5,665 persons for the NSHA study group.

The Pilot Study[3]

In response to a recommendation in the committee's first interim report (IOM, 1997a), plans were developed for a large comprehensive pilot study preceded by extensive testing before the conduct of the national study. Extensive plans for testing were developed for the pilot study. These included a comprehensive series of tests and experiments covering all aspects of the survey operations, design, response rates, and the content and effectiveness of the questionnaires before the start of, and during, the pilot study. A sample of approximately 13,200 households was expected to be contacted in eight PSUs in the initial screener.

The purposes of the pilot study were to experiment with several data collection methods and procedures, and to ensure that the questionnaires were clear and concise, that all procedures ran smoothly and efficiently, and that the burden and discomfort placed on the respondent were kept to a minimum. Other purposes included testing the effectiveness of the screening instruments and measuring the accuracy of the screening algorithm; evaluating procedures to maximize response rates—both total and item response; and developing estimates of prevalence rates to determine the final sample sizes for the main study. Finally, the pilot study was also designed to test the operational procedures for medical examinations, including measuring the reliability of physician and nurse practitioner examinations; to evaluate medical examinations performed in the home and in mobile examination centers (MECs); and to measure the reliability and validity of the simulated disability decision process. The pilot study was also designed to test instrument designs for the DES and more thoroughly test the screens and questionnaires. The tests concerned the screener methods used to allocate the general population into the four study groups.

The Time Demands to Achieve Survey Quality

A major lesson learned from the experience in planning and developing NSHA is that before starting a national survey, sufficient time should be allowed to (1) conduct and analyze the results of the various pretests,

---

[3]For a more detailed description and discussion of plans for the pilot study, including instruments, procedures, design, and response rate goals, the reader is referred to Westat, 1999a,b,c,d, and IOM, 1999b.

focus groups, and cognitive tests; (2) conduct a comprehensive pilot study with the planned and other built-in experiments; and (3) analyze and test alternative solutions in areas that need resolution as a result of the pilot study.

NSHA implements survey measurement of complex concepts in the absence of a scientific consensus on what measures are best suited. It is on the frontiers of survey design. When survey measurements must be crafted without the benefit of years of prior development, great care must be taken in assessing whether they measure what is intended. Similarly, screening protocols and physical measurements require time for development and evaluation prior to their use in production settings.

Because of significant committee uncertainty about the effectiveness of the survey instruments to measure disability, the committee strongly recommended in its interim report (IOM, 1997a) that SSA set aside a significant amount of time and resources for NSHA questionnaire design and testing. The committee also recommended a rigorously designed field experimentation and development phase of the survey to identify mechanisms for enhancing participation in the survey, to establish the validity of measures obtained, to assist in the quality of medical records obtained, and to guide decisions on issues relating to medical examinations. The rush to launch the national survey, however, caused serious logistical inflexibility during the various phases of the survey.

The pilot study is an example of allowing inadequate time for the development and testing that is required. SSA planned to complete developmental work and conduct a pilot six to nine months after award of the contract for the survey. Following the committee's recommendations, SSA developed extensive plans for a large comprehensive pilot study including testing all exploratory information and procedures through focus groups, cognitive laboratory tests, and pretests.

The pilot study was conducted in the first half of 2000 with about 12,000 initially selected households and a completed database of nearly 4,000 cases. It was conducted in four counties (and not eight as previously planned) selected for their geographic and regional diversity. Only a short period of time was allowed in the schedule for development, testing, and making the necessary modifications before launching the national survey. Decisions had to be made throughout the process, and the results of the pilot study made it obvious that there was insufficient time to resolve issues and test alternatives before launching the national survey.

Several reports evaluating the results of the pilot study were prepared by Westat identifying corrective revisions made during and immediately following the pilot study, and recommendations to SSA for further revisions that would be tested before implementation in the main survey. The revisions were focused on achieving two goals: (1) reducing the burden on respondents and (2) maximizing the capacity of the items to produce data needed to answer the research questions posed by SSA. The revisions, therefore, took on an iterative process aiming to strike a balance between these two goals, at times with a possible net result of no reduction in respondent burden. Several small-scale pretests were planned, some already were under way or had been completed at the time of this writing. These pretests should provide feedback on instrument length, flow, item clarity, and item sensitivity.

As a result of the pilot study experience, the data collection plans are being restructured, and the mode of data collection changed because of poor results with the random digit dialing (RDD) sampling frame. Westat will be using area sampling and will try to get telephone numbers for the sampled persons. If successful in obtaining telephone numbers, the screening interview will be conducted by telephone. If unsuccessful, a field interview will be administered. Westat expects to get about 25 percent of the responses by telephone. The screening interview also is being revised with the goal of reducing the respondent burden to about 20 minutes.

One of the primary concerns expressed by the DDSs was that the information presented to them in the NSHA data packet from the pilot study did not always seem complete. This led to their lack of confidence in making a simulated disability determination prior to the full survey. SSA and Westat are planning to conduct a small "end-to-end" test involving about 100 persons, most all with known disability status. The main purpose of this test is to check that the revisions made to the data collection procedures do in fact improve the completeness of the data collected on respondents to determine medical and vocational eligibility for SSA benefits.

The committee in its third interim report had concluded that it seriously doubted that enough time was allotted to determine what changes are needed and to implement those changes before the conduct of the national survey. In order to assess the findings of the pilot study and resolve the problem areas in a satisfactory manner, more time will be needed between the completion of the pilot study and the start of the national study than the two to three months allocated. The time frame provided little flexibility in terms of the amount of time available to make deliberate and rigorous decisions on issues of design, procedures, and questionnaire if problems are uncovered during the pilot study. The committee recommended that SSA revise the project schedule to allow significantly more time to plan and analyze the pilot study and test alternative solutions for problem areas before starting the national study. Unless the period for testing, analysis, and development is extended, SSA could encounter serious problems during the national survey. The committee recognizes that increasing the time and level of research between the pilot study and the national survey may have cost implications. The committee understands SSA and Westat are already addressing many of the issues raised in this report. The committee notes that since then, SSA has approved significant additional time to the schedule to adequately evaluate the results of the pilot study and to test alternative solutions for problem areas before starting the national study.

Given the complexity of the NSHA, the committee in its interim report (IOM, 1999b) also suggested the conduct of a dress rehearsal once all the issues are resolved and before starting the national study. No time had been allocated for a dress rehearsal in the timetable for the study. In response to the committee's recommendation, however, a dress rehearsal is included and will be the last step before nationally representative data are collected in the main survey. It is slated to begin only slightly ahead of data collection in the first year of the main survey. Preliminary work on the dress rehearsal is expected to begin March 2002. The actual interviews and examinations will be conducted between December 2002 and January 2003. As of July 2001, plans called for the field work for the main survey to be carried out over multiple years beginning in early 2003. The full NSHA sample of 80 PSUs will be divided into two or more replicates, each of which will be nationally representative. This design will provide the ability to assess response rates and the ability to obtain preliminary estimates at the end of the first replicate.

In summary, not allocating sufficient time in the beginning for research, development, and testing prior to launching a major complex survey has resulted in the need to repeatedly revise the timetable for the various steps in the development and conduct of the survey. To illustrate: the original schedule for planning, development, and completion of the survey as reflected in SSA's request for proposals for contract covered a total of two and a half years from January 1998 to August 2000. Ten months were allowed for the award of the contract, planning, and development, and 10 days later a pilot study was planned, with no time for iterative testing and experimentation before the pilot and between the pilot and the start of stage one of the survey. In response to the committee's concerns and recommendations issued in its first interim report (IOM, 1997a), the pilot study was delayed, but only by about a month. SSA also assumed that all analysis and revisions could be done during the pilot study and so allowed only two to three months from the end of the pilot study (November 2000) to the start of the main survey (January 2001); therefore very limited time was allowed for research, development, testing, and making the needed changes. Although some decisions on instrumentation can be made prior to the end of the pilot study, a thorough analysis of issues was not possible until the end of the data collection phase in the pilot study. Even if analysis of some tests and experiments could have begun earlier in the analysis phase of the NSHA pilot study, additional time would have been needed to examine the implications and plausibility of several different "adjustments" in the problem areas. As indicated earlier in this chapter, the results of the pilot study made it clear that revisions and more iterative testing of the revisions were needed. The most recently revised schedule available to the committee called for the end-to-end test data collection from December 2001 to February 2002; dress rehearsal data collection from December 2002 to January 2003; and the main study to start early in 2003. Thus, the survey that was originally planned for the middle of 2000 is now scheduled to start in 2003 and assumes a five-year data collection plan.

NEEDED RESEARCH IN THE MEASUREMENT OF DISABILITY IN A SURVEY CONTEXT

The experience to date with the NSHA, as well as work with other surveys that include measurement of disability, makes clear that the measurement of people with work disabilities is complex. The complexity stems, in part, from differences in conceptual models of the enablement–disablement process and alternative interpretations of the various conceptual models discussed in Chapter 3. In addition, there exists an incongruity between the various conceptual models and SSA's statutory definition of work disability. The various constructs do not necessarily identify the same population. Finally, NSHA must address both the estimation of how many persons might apply for SSA benefits and the number that would be classified as persons with work disabilities in the SSA benefits decision process.

All complex surveys such as the NSHA require trade-offs between the cost of the survey, the timeliness of the survey statistics, and the quality of the statistics derived from the survey. For example, quickly mounted surveys, especially in new fields, can rarely produce high-quality statistics, although they may save the sponsor money. Quality in survey statistics, in turn, has a well-established structure in surveys, involving closeness of the responses obtained to the true underlying attributes of sample persons, on the one hand, and the ability of the resulting set of respondents to represent the characteristics of the full U.S. population, on the other hand.

Although a number of research activities are under way worldwide that address issues related to statistical error associated with the measurement of disability, these efforts are but a beginning with respect to understanding the properties of measurement error associated with disability-related questions. In addition, other sources of error are, for the most part, not addressed in current research activities. The committee sponsored a workshop in May 1999 to bring together disability researchers and experts in survey methods to discuss conceptual and survey design and measurement issues, and to identify unanswered questions of measurement of persons with work disabilities (IOM, 2000). The discussion revealed several gaps in survey methods and measurement of work disability, leading to a framework for long-term research for SSA and others in the field. This framework encompassed four broad areas of research, paralleling the stages of survey measurement: (1) coverage error, (2) measurement error, (3) nonresponse error, and (4) the development of measures of the environment. Each of these areas is discussed briefly below, with specific references to NSHA.

Coverage Error

Coverage error is produced by the failure to include all eligible people on the list or frame used for identifying and sampling the population of interest. The use of screening questions to identify the population of interest leads to an additional source of coverage error—the exclusion of persons due to inaccurate classification at the time of the screening.

Household-Based Surveys

Household-based surveys by definition eliminate from the sampling frame those members of the population who are homeless, as well as those living in institutions. Those residing in group homes, assisted-living facilities, and other new types of residences may or may not be included in the frame, depending on how the distinction is made between institutional and noninstitutional residence. SSA likewise has decided to exclude from the NSHA the institutionalized population and the segment of the homeless population who cannot be found in households or other quarters at the time of the interview.

However, the question of including or excluding homeless people in the NSHA is not as straightforward as the other household surveys. (The committee discussed the issues surrounding the inclusion of the homeless and institutionalized population in its interim report; IOM, 1997a). The committee recognizes the likelihood of relatively high rates of disability among homeless and institutionalized populations, and the resulting negative bias resulting from their exclusion. The extent of this coverage error, when attempting to describe the entire U.S. population with respect to disabilities, is unknown. It is likely to be a function of the type of disability, with estimates of the population with mental retardation or mental health problems most likely subject to the highest rates of coverage error. Empirical data are needed to estimate the differences in the rate and characteristics of the population with disabilities based on household surveys as compared to the entire population.

At the same time the committee has serious questions about the operational and methods issues involved in attempting to include homeless and institutionalized populations in NSHA. Can reliable information be obtained, feasibly and economically, from homeless and institutionalized populations? Techniques have been developed to locate, sample, and obtain data about each of these populations. Yet locating and screening respondents for eligibility require special efforts involving careful, and long-term planning, large amount of staff resources, considerable time, and high levels of funding. Homeless people present problems in scheduling, interviewing, and administering performance tests and medical examinations. Maintaining contact with them and getting them to participate in adequate numbers in the medical examination also would be problematic. Likewise, obtaining permission from family members for the participation of people in long-term care institutions who are not able to grant permission themselves may be difficult.

The committee concurred with SSA that adding homeless and institutionalized populations to the sampling frame at this time would not be cost-effective. Much research and testing are required to develop the necessary protocols and procedures for conducting the NSHA among homeless people and those living in different types of institutions. The costs of sampling and interviewing in the various types of institutions would be prohibitive. Thus, limiting the target population to the household population seems appropriate. In its earlier report the committee urged SSA to undertake research as part of its long-term research plan leading to the inclusion of these populations in subsequent studies or a separate supplement to future surveys such as the NSHA.

Effects of Alternative Approaches to Screening

The use of a screening instrument to identify the population of interest often impacts coverage error. The committee believes that three areas of research are particularly important with respect to the use of screening instruments:

1. the effect of alternative wording of questions on the identification of the population—given the discrepancy among rates of disability evident in the literature, establishing the reliability of screening items is particularly important,
2. comparisons of estimates based on simultaneous screening and interviewing with those based on separate screening operations—

this research should also focus on understanding the mechanism by which the two operations result in different estimates, and

3. the effect on estimates when a subsample of cases classified as negative according to screening questions is included and re-screened as part of the extended interview (this approach is taken by Statistics Canada in its Health and Activity Limitations Survey).

SSA in its survey plans had specified the use of telephone number frames for NSHA. Households with telephones were to be selected by list-assisted RDD sampling. This decision by SSA appeared to be driven primarily by cost considerations. The choice of sampling frame determines the nature of noncoverage error in any survey. Common choices in surveys in the United States are area frames, offering theoretically complete coverage of households and institutions; dual-frame designs combining telephone and area frames; dual-frame designs combining area and institutional list frames; and telephone number frames.

The committee expressed serious concerns about the adequacy of coverage of the general population based on RDD sampling. Noncoverage of persons in households with no telephones should be of particular concern for persons with disabilities. In addition, there was no indication of how SSA will deal with people with hearing loss, communication disorders, mental and cognitive impairments, and emotional disturbances, who are not likely to be covered well in a household frame.

Approximately 5 percent of households in the United States are without telephones. Moreover, persons in households without telephones have a higher rate of disability (17 percent) than those in households with telephones (15 percent) (Thornberry and Massey, 1988; LaPlante and Carlson, 1996). The availability of telephones also is negatively correlated with income.

In addition, telephone sampling and screening would likely offer lower response rates than face-to-face screening (Groves, 1989; Lessler and Kalsbeek, 1992). As a consequence the screening sample would need to be increased to compensate for the losses from the sample because of nonresponse; the higher nonresponse rates are likely to increase the risk of bias in the estimates. Thus, although telephone screening may be less expensive, some aspects of the quality of the data collected are more suspect. Careful study of mechanisms to increase the screener response rate is required. These mechanisms might include incentives, refusal conversion efforts, switches to alternative modes of data collection, and so on.

Also, there was no indication by SSA how it would deal with people with hearing loss, communication disorders, mental and cognitive impairments, and emotional disturbances. SSA also has the problem of response burden for the total household if more than one person in the household has a disability and proxy reporting is not encouraged. Similar problems will have to be faced in the main interview and in administering medical examinations and performance tests to persons with severe disabilities. The effect on response rates and bias could be significant. The committee advised in its interim report (IOM, 1997a) that SSA should test several options dealing with these problems in pretests prior to the start of the national survey.

In terms of coverage of the adult working age population, survey response rates, and some features of the screening measurement, the preferred design is an area probability, face-to-face survey. It is also clear that the cost of such a design would be higher than the alternative proposed by SSA. The additional costs for a survey of this importance and complexity should be considered in the context of the size of the program itself (SSDI and SSI) and the implications of poor or imprecise information. The committee, therefore, urged a careful review of the costs of a full area probability survey, in light of the cost savings proposed in later recommendations.

These concerns about the exclusion of non-telephone households led the committee to recommend in its first interim report that NSHA should be based on a design offering full coverage of the U.S. household population of adults. The committee recognized that the cost of including persons in non-telephone households would increase the costs of NSHA. The committee therefore recommended that if resources were lacking to use an area probability sample using face-to-face interviews, the Social Security Administration should use a multiple-frame design of a statistically optimum mix of RDD and area frame of the general population followed by face-to-face interviews of the eligible population.

The NSHA pilot study demonstrated that while the cost of using a sample from the RDD frame was lower than that of an area frame, the resulting response rates (a risk indicator for nonresponse error, reviewed below) were much lower. After the pilot, consistent with the committee's earlier recommendation, Westat has recommended to SSA that an area frame design be used, offering greater coverage of the household population and likely better response rates, at likely higher costs.

Proxy Respondents

The issue of the use of proxies arises in this survey because a large number of people in the sample will have disabilities or some kind of functional limitation. Westat plans to avoid proxies whenever possible. However, it may be necessary to collect information from proxies to ensure the highest possible response rate and to obtain as much information as possible from people who have difficulty responding on their own.

Westat's plans call for a household reporter to answer questions in the initial screener about all working age adults in the household. Westat is concerned, however, that such reporters may not be able to answer accurately and honestly questions about the mental and cognitive health of other members of the household. Westat is also concerned about the risk of very low response rates if it attempts to interview each person in the household about his or her mental and cognitive health. During the follow-up screener and the comprehensive survey interview, Westat plans to use medical exam proxy assistants in interpreting for and assisting the sample person with medical needs or language problems (Westat, 1999c).

Proxy interviews have varying levels of accuracy depending on the topic of the interview and the relationship of the subject to the proxy. Westat believes that the use of proxies in the initial screening process will make it oversensitive; for purposes of the initial screener, however, that would be acceptable. Beyond the initial screener, Westat plans to avoid using proxy reporters but does expect to have proxy-assisted interviews. The decision to use or not use a proxy respondent will be made when the sample person is initially contacted. If the respondent is available and able to complete the interview, the interviewers will be discouraged from accepting a proxy (IOM, 1999b; Westat, 1999c).

The committee believes that the issue of proxy respondents is an area for fruitful research as noted below.

Sampling Error

Most users of survey data know that larger samples reduce the uncertainty that the survey results will depart from those in the full target population because of the subset of the population that was sampled. Sampling error can also be reduced by stratification of the frame into separate diverse populations, followed by independent selections from each subpopulation or stratum. Conversely, use of clustered samples (e.g., sampling persons together who live in the same geographical area) and assignment of vastly different probabilities of selection can increase the instability of survey statistics due to sampling error. NSHA samples will have to be clustered given the use of the MECs to conduct the medical examinations and tests.

SSA assumed that the core group sample of 3,090 will be sufficient to estimate several subgroups of particular policy interest. These subgroups include potentially eligible nonbeneficiaries who are working; younger nonbeneficiaries with disabilities; nonbeneficiaries aged 62–69 years;

nonbeneficiaries with mental, emotional, or behavioral conditions; and nonbeneficiaries with disabilities from minority groups.

The committee expressed concerns in its interim reports about the adequacy of the size of the total sample and of the allocations among the four subgroups—nonbeneficiaries with severe disabilities, persons with significant but lesser impairments, nondisabled persons, and current beneficiaries—and questioned SSA about this disproportionate sample design and the basis for choosing the specific sample sizes for the four groups. The committee could not understand the logic that led to this particular disproportionate sample design. It believes that the targeted sample sizes would lack the condition specificity that SSA would require for estimation and analytical purposes. Even if SSA can achieve these planned sample sizes, the cells very likely will be much too small, especially if SSA stratifies on more than one disabling condition and/or demographic or socioeconomic characteristics such as age, gender, minority status, or working nonbeneficiaries with specific disabling conditions.

Similarly, the proposed sample size for the borderline group of persons with less severe disabilities may not be sufficient in its analytical strength for assessing how alternative decisions and policies would affect outcomes. The differences in outcomes resulting from changes in policies or procedures is likely to be minimal, if any, for persons with severe disabilities, but some real differences could show up among borderline cases under alternative conditions.

The committee expressed similar concerns in its third interim report and continues to have several questions and concerns about the adequacy of the total sample size and especially about the allocation of people among the four subgroups. The sample sizes may not support SSA's requirements for estimation and analytical purposes. As stated above, the committee does not understand the logic that led to these sample sizes and allocations. It has not seen the statistical rationale for setting the sample size targets or the plans for analysis that would drive the sample and content of the survey.

Nonresponse Error

Although adequate empirical data do not exist to measure the impact of nonresponse on estimates of persons with disabilities, the nature of a person's impairments or disabilities might result easily in differential nonresponse among members of the population with disabilities. This deficit in the literature suggests that a priority for nonresponse research is the assessment of differential nonresponse among persons with disabilities.

The role of gatekeepers and interviewers may represent sources of nonresponse error unique to the measurement of persons with disabilities. Gatekeepers may limit access to persons with disabilities who, if provided with the opportunity, might be quite willing to serve as respondents. The role of gatekeepers, their contribution to nonresponse, and the differential impact of gatekeepers for telephone surveys compared to face-to-face administration of interviews have never been addressed in the literature. Similarly, interviewers may classify sampled persons as incapable of serving as respondents, due to apparent cognitive, sensory, or other impairments. Research also is needed to address the extent to which such judgments by an interviewer result in nonresponse among the population of primary interest.

SSA had assumed that at response rates of 90 percent for each component of the NSHA, it should get the planned sample sizes. The committee repeatedly has stated that the expected rates may be overly optimistic, especially for a population with disabilities. It raised these issues in its first interim report (IOM, 1997a); it reemphasized in its third interim report (IOM, 1999b) the problems that could arise as a result of sample selection, size, and allocation if adequate advance planning and testing are not undertaken.

The committee has learned recently that SSA is rethinking these targets. As a result of experience with the pilot study, SSA has reevaluated the response rates and now believes that response rates of 85 percent for the screening interview; 85 percent for the in-person interview; 90 percent for the medical examination; and an overall response rate of about 60 percent are more realistic to achieve. SSA also is now revising upward the sample size estimates on the basis of information from a number of sources including the simulation experience from the pilot study. This process will not be finished until the "end-to-end" test is completed. (Personal communication, John R. Kearney, Office of Research, Evaluation, and Statistics, SSA, March 21, 2002.)

Respondent Burden

Each of the NSHA survey instruments used in the pilot is lengthy and complex, thus creating a risk that respondents will be unwilling or unable to provide useful data to SSA. For example, SSA has noted that the Comprehensive Survey Interview will impose a burden on some respondents who have a complicated medical history, considerable income or assets, and a complex work history. The committee agreed and expected that other NSHA components will also impose a significant burden on these and other respondents. Another concern is the initial screener, because its results will be used to sort individuals into the four categories. For this screener, one household member will be asked to respond to numerous questions, including questions about mental and emotional problems, for all household members 18–69 years of age. If the informant does not answer these questions correctly for all household members, individuals who have conditions that should result in their selection for the follow-up screener may be missed.

Because of its length and complexity, SSA and the committee agreed that the instrument would have to be reduced in length between the end of the pilot study and the start of the national study. SSA first must decide which questionnaire items are to be eliminated, and then the shortened version must be evaluated and field-tested to ensure its viability as an instrument that can meet the study's goals. These steps will take several weeks or months to be done well. In its third interim report the committee recommended that SSA revise the project schedule to allow significantly more time to plan and analyze the pilot study and test alternative solutions for problem areas before starting the national survey (IOM, 1999b).

Measurement Error

Estimates of the population appear to vary as a function of the essential survey conditions under which the data are collected, specifically, the mode of data collection, the wording of the specific question, the context of the question, the overall content of the survey, the survey's sponsorship, and the nature of the respondent providing the information (self versus proxy response).

Regardless of the type of impairment, the development of valid and reliable measures of disability—especially work disability—is a challenging undertaking, but their episodic nature, as well as perceptions of social stigma make the measurement of mental and cognitive impairments all the more difficult. Valid and reliable measures of participation in the social and economic environment are needed. Valid questions should reflect the conceptual models that view work disability as a matter of degree, suggesting that the measurement of disability be on a continuum as opposed to the dichotomous measures used in many surveys.

Three areas of research are needed for developing valid and reliable measures of work disability:

1. *Assessment of the effects of specific question wording and question context.* This involves

   • research directed toward understanding respondent's comprehension of the key concepts within the question, such as "difficulty," "work," "performance," and "ability";

- decomposing long questions used to screen for persons with disabilities and making comparisons between the approaches with respect to reliability, validity, and length of administration; and
- assessment of the role of context on estimates of the population where context is broadly defined, ranging from subjective factors such as mood to objective factors such as the survey sponsor, the questions immediately preceding the disability measures, and even such factors as the weather.

2. *Assessment of the effects of self and proxy reporting:* A limited empirical literature on the effects of self and proxy reporting of functional limitations suggests that the direction and magnitude of response error is, in part, related to whether the report is provided by the individual or by proxy. (See for example LaPlante and Carlson, 1996; Todorov and Kirchner, 2000.)

3. *Assessment of the effects of essential survey design features:* Estimates of persons with disabilities or persons with work disabilities vary as a function of essential survey design features. Some examples of design features include sponsorship of the survey that could affect both the properties of nonresponse (motivation to respond or not respond) and the measurement process (response editing and formation); the effects of the presence of others during a survey administration, especially in the measurement of mental illness; the effects of mode of interview; and incorporation of new technology (e.g., audio computer-assisted interviewing) to enhance participation and privacy among persons with disabilities.

The Challenge of Measuring the Environment

One of the major voids between conceptual models of impairment and disability and survey measures is the inadequacy of survey questions to measure the environment. Current data collection efforts, for the most part, fail to measure the environment and its impact, either as a means of facilitating or as a barrier to participation in the social and economic environment.

Environmental factors are external factors that make up the physical, social, and attitudinal environment in which people live (Fougeyrollas, 1995; Friedman and Wachs, 1999; Schneider, 2001; Whiteneck, 2001). The classification of environmental features enumerated in the second revision of the International Classification of Functioning, Disability and Health (ICF), (formerly the International Classification of Impairments, Disabilities, and Handicaps [ICIDH]) provides a well-defined architecture for developing questionnaire items designed to capture environmental factors that affect the disablement process. Among the environmental factors of importance in the ICF framework are products and technology, the natural environment, support and relationships, attitudes, social services, systems, and policies. Of interest with respect to disability is the extent to which environmental factors either facilitate or present barriers to participation in social roles. As part of the research to design questionnaires that map conceptually to the ICF coding framework, researchers are currently addressing the development of both objective and subjective environmental measures (Schneider, 2001).

The committee underscores the need to develop measures of both the physical and the social environments. The measurement of environmental context should examine both factors that accommodate impairments and those that serve as barriers. The development of objective measures of the physical environment may be facilitated by fostering collaboration with researchers in ergonomics and human factors engineering, fields in which a primary focus is the measurement of the environment.

To aid in the development of objective measures of the social environment, the committee notes the need to develop and test questions concerning social climate, barriers, and stigma. These questions are especially important for those with mental illness, but they are relevant for, and should be asked of, all persons with disabilities.

One of the challenges related to developing objective measures of the environment is the identification of a set of questions that can be asked of the general population. However, to fully understand either barriers to employment or factors that facilitate employment, questions must be tailored so as to be relevant to the individual's situation. Ethnographic exploratory studies of workplace environments are one means by which to inform household measurement of accommodation and barriers. For those who are no longer working, questions that enumerate what accommodations would be necessary to facilitate, or what barriers prevent, participation in the workforce have to be designed and subjected to evaluation. Similarly, research is needed on developing subjective measures of both the physical and the social environments that either facilitate or limit participation.

In addition to research for developing such measures of the environment, research also is needed on two additional topics: (1) assessment of systematic differences in evaluating the environment among those for whom the environment is benign versus those for whom the environment is hostile and (2) assessment of the difference between self and proxy subjective reports of environmental conditions.

To summarize, the empirical literature examining measurement error associated with specific questions, albeit limited, suggests that items currently used to screen or measure persons with disability are subject to low levels of reliability and are of questionable validity. The impact of both coverage error and survey nonresponse on estimates of the population with disabilities and work disabilities has not been addressed in the literature. In light of these points, the measurement of people with disabilities as well as work disabilities could be greatly improved with research directed toward one or more of these agenda topics.

Although a number of research activities are under way in the federal agencies (Hale, 2001; Rand, 2001) that address issues related to response validity and reliability associated with the measurement of disability, these efforts are only a beginning with respect to understanding the properties of measurement error associated with disability-related questions. Other sources of error identified above—most notably coverage and nonresponse error—are for the most part not addressed in current research activities. Without an understanding of the extent to which coverage error and nonresponse error impact estimates of work disability, it will be difficult for SSA to monitor the size and characteristics of the potential pool of applicants based on survey data. *SSA, in collaboration with other federal agencies, should engage in an ongoing program of research on measurement issues, taking into consideration the conceptual developments in the field.*

The impact of the research efforts designed to address measurement error on subsequent rounds of NSHA and related data collection activities is that in the near and intermediate future, questionnaires incorporating measures of disability will be in a dynamic state. Changes to question wording and response options are likely as research reveals the characteristics of questions and design features that result in higher-quality (validity and reliability) measures of disability. Question wording identified in the current NSHA for monitoring the pool of potential applicants for disability benefits, or models using questions in current use, may be obsolete in the near future, as surveys adopt new questions or design features to minimize response error.

Because SSA had not mounted an ongoing program of survey measurement of disability for many years, much of what it is attempting in NSHA is novel. New survey measurement demands careful, time-consuming development. For measurement involving questions, qualitative research probing issues of comprehension by diverse respondent groups is needed. Cognitive interviewing techniques are used to examine the memory structure of respondents relevant to the material being measured. Computer-assisted interviewing software needs to be designed to improve memory cues and reduce psychological threats to measurement error. The reduction of survey nonresponse requires that interviewers identify and address the concerns of different types of sample persons to the survey request. Finally, all the components of the survey must be tested together in a pilot study or dress rehearsal.

Such research, when conducted extramurally, but guided by the mission of the agency, can provide the agency with proven measurement approaches when new concepts become integrated into statutes guiding program designs. For example, the Disability Research Institute (DRI) established by SSA in May 2000 could serve as a useful vehicle for the conduct of the research discussed above.

FUTURE SURVEYS OF DISABILITY AND WORK

The enduring lesson of the NSHA for other survey efforts to be undertaken by SSA as part of the work disability program is clear—careful survey design and measurement require considerable development and field-testing prior to implementation. Cost savings that appear to arise when work is rushed are illusory. Cutting corners can be done only with careful, experience-based judgments and analysis. Delays in the original schedule of the NSHA that evolved over the course of the committee's interaction with SSA often arose because unanticipated discoveries were made about the complexity of the survey design and implementation tasks. It is likely that the total cost and total time of the project are greater than would have occurred if more careful, deliberate developmental studies had preceded the launch of the major national survey.

The committee has repeatedly stated during the course of the study and in its interim reports that the NSHA, if well designed, could be the cornerstone for long-term disability research. When completed it can be of fundamental importance to future analyses by the SSA and other researchers. It will provide information that would guide SSA in making decisions about its disability programs and will play a key role in projecting and understanding disability rolls in the future. Moreover, it will lay the groundwork for future surveys. *Early in the study the committee strongly endorsed the conduct by SSA of a well-designed, carefully pretested, and statistically sound survey. The committee reiterated its position later in the study. It has not changed its position today.* Rather it reemphasizes its endorsement. However, the value of the information diminishes with time. It is therefore critical that SSA update the comprehensive database with regular periodicity. To ensure effective planning, SSA must examine the fundamental characteristics of who has work disabilities, and how many more, or fewer, people will become eligible. SSA has not collected such information for more than 20 years, and it is long overdue. It is critically important that SSA not wait another 20 or more years before obtaining such basic information so relevant to its policies and programs.

Recommendation 4-1: The committee recommends that prior to undertaking any future large-scale data collection effort, the Social Security Administration should allow sufficient time and provide adequate resources to

- **investigate, test, and incorporate conceptual developments; and**
- **develop, pretest, pilot test, and revise measurement instruments and design.**

In conclusion, the immediate need of the NSHA involves estimates of the size and characteristics of the pool of persons eligible for SSA disability benefits. A cross-sectional sample of the household population done at a particular point in time provides useful estimates for such needs. When change over time is an issue, survey measurements must be repeated in order to provide estimates of change. When the only interest is whether the full target population has experienced a change in the prevalence of a phenomenon, an independent cross-sectional survey conducted at a later time provides useful change estimates. When the interest concerns whether some types of individuals change and others do not, a longitudinal survey, conducting repeated interviews of the same persons, provides the most useful data.

SSA's needs for the estimation of change over time in the size and characteristics of the eligible population stem from the necessity to forecast the growth or decline of the applicant and beneficiary pool. SSA has stated that NSHA will permit forecasting of changes in the size of the beneficiary population. Such a goal implies ongoing measurement of the size and characteristics of the eligible population, with updated instrumentation to reflect any changes in conceptual and measurement issues and in SSA's eligibility protocol that may have occurred in the intervening years.

The next chapter discusses the design choices for obtaining the needed information on an ongoing basis using a reduced set of measures in the intervening years between the conduct of the large surveys.

5

A Work Disability Monitoring System

The previous chapter discussed the National Study of Health and Activity (NSHA) developed by the Social Security Administration (SSA) to estimate the size and characteristics of the population eligible for disability benefits. When completed it should yield a rich set of data that should be valuable for policy development and planning. Effective management of SSA's disability programs requires sufficient information to understand and predict changes in the size, characteristics, and distribution of the pool of persons eligible for disability benefits (applicants and beneficiaries), as well as to understand the factors that affect application volume and answer many policy questions. A single cross-sectional survey such as the NSHA will not provide adequate data in the future for either of these goals. Medical models of disability historically have been insufficient in explaining unexpected growth in the size of the applicant pool (Haber, 1971; Yelin et al., 1980; Stapleton and Dietrich, 1995; Stapleton et al., 1995; Bound and Waidmann, 2000). Factors extrinsic to the benefits programs—for example, cyclical changes in the economy, as well as social and cultural issues—have resulted, in the past, in changes in applications and awards rates and unexpected increases in program expenditures.

One means to understand the magnitude and characteristics of the potential eligible population as well as the intrinsic and extrinsic factors that influence the application for benefits is to develop a monitoring system related to work disabilities. The idea is not new, as is evident in the series of surveys sponsored by SSA throughout the late 1960s and the 1970s (for details, see Mathiowetz, 2001, in Part II). Of interest in a disability monitoring system is not simply measurement of the prevalence and of the socioeconomic conditions linked to disability, but also understanding both the individual and the environmental factors that lead to changes in the application process. This chapter explores ways in which SSA could build from its experience with the NSHA to develop an ongoing disability monitoring system for Social Security programs to provide timely information on the prevalence and distribution of disability in the working population. This chapter discusses the need for such a system, essential principles of such a system, possible design choices, and a suggested planning and implementation strategy.

NEED FOR A WORK DISABILITY MONITORING SYSTEM

A well-designed monitoring system should provide SSA with the data needed to respond to a variety of policy and planning issues, including, but not necessarily limited to, the following:

a. Size, distribution, and characteristics of the working populations with disabilities. The growth in the population eligible for SSDI and SSI during the past three decades and the concomitant growth in applicants and awards have raised questions as to whether continued expansion of these programs can and should be sustained.
b. Demographic trends. The working age population has grown dramatically and its composition has undergone fundamental change since the inception of the SSDI and SSI programs. This working age population eligible for disability benefits is projected to increase in the coming years as the baby boom generation ages and reaches the ages at which chronic diseases and disabilities are more likely to occur. This growth will impact significantly the Social Security disability programs in many ways.
c. Labor market dynamics. Structural changes in the economy such as the relative shift over the years to service industries and occupations have a significant impact on the types of impairments that result in work disability. Labor force participation rates among women have increased substantially while those of men have declined. These structural changes need to be fully understood and predicted accurately.
d. Changes in economic conditions. During periods of slowdown in the economy and high unemployment, marginal workers especially low-wage workers with disabilities are more likely to apply for disability benefits. SSA needs to closely monitor these changes in economic conditions and their impact on Social Security disability programs.

e. Needs of minority and special populations with disabilities. Disabilities are disproportionately represented among minorities, the elderly, and lower socioeconomic populations. The causes of these differentials are not clear. People with lower socioeconomic status probably experience more injuries, higher mortality rates, less access to health care, and generally poor health. On the other hand, some people have lower incomes because their disabling conditions restrict their ability to work. These phenomena and their relationship to application and receipt of Social Security benefits need to be further studied.

f. Quality of life for disabled workers. Quality of life is an important theme for all workers, but it is especially important for those with disabilities. An improved quality of life through provision of assistive technology in and out of the workplace could represent the difference between working and applying for Social Security benefits. Quality-of-life measures for the population with disabilities need to be developed by SSA in collaboration with other relevant agencies.

g. Functional status. Health conditions differ in the degree to which they cause functional limitations and disabilities that may result in work disability. One of the committee tasks was to examine SSA's research into functional assessment instruments for its redesign efforts and to provide advice for adopting or developing instruments for the redesigned decision process and the National Study on Health and Activity. A workshop titled Measuring Functional Capacity and Work Requirements was held on June 4–5, 1998. Following the workshop, the committee issued its second interim report in 1988 titled *The Social Security Administration's Disability Decision Process: A Framework for Research.* The committee recommended that further research on functional assessment measures be conducted. The role of functional assessment in the disability decision process still remains an important issue.

h. Role of the states. As noted in Chapter 2, in times of poor economy, cuts made in state and locally funded general assistance and other welfare programs result in shifting the burden from state and local programs to federal programs. In addition, welfare agencies routinely refer persons to SSA's disability programs. The relationship between SSA's disability programs and state and locally funded programs needs to be further studied.

i. Legislative, regulatory, and judicial impacts. Legislative and regulatory changes and court decisions have a major impact on SSA's disability programs. These need to be monitored on an ongoing basis.

As stated in the previous chapter, SSA considers NSHA the center-piece of its long-term disability research to understand the growth in the disability programs. The NSHA is designed to provide SSA with an estimate of the extent of the prevalence of disability, the factors that enable some people with disability to remain in the workforce, a basis for gauging the effect of changes in disability decision criteria, and much needed insight into the problems of measuring disability in surveys. *The committee believes that the NSHA should be considered the first important part of a long-term commitment by the SSA to produce reliable national data on the demand for and quality of its disability benefits programs.*

However, NSHA will be of limited utility in directly measuring long-term temporal market changes and demographic changes. Although the population may age in somewhat predictable ways as the baby boom generation moves into the vulnerable ages, the size and geographic distribution of the racial and ethnic makeup of the population will change in less predictable ways depending on the swirling currents of economic opportunity and the associated flow of immigration across the nation's borders. Uncertainty about the future direction of legal and policy changes affecting the population with disabilities, and about ways in which medicine and technology might be used to enable Americans with disabilities to function more effectively, will also limit the long-term usefulness of NSHA data. The committee believes that SSA has a continuing need for current and reliable data to project growth in its programs, and to understand the contributing factors. Similarly, it needs data to effectively adapt its disability benefits programs to the changing needs of people with disabilities.

Disability is a dynamic phenomenon that needs to be monitored and evaluated continuously. The conduct of in-depth periodic disability surveys will provide an essential database for understanding this dynamic process. Because of the time lag for research and development, as well as the costs involved in launching a survey of the magnitude of the NSHA, it is not feasible to repeat such a study design every year, or even every other year. *Such a comprehensive survey should be conducted with regular periodicity, at least every 7–10 years. In the intervening years, however, SSA requires ongoing estimation of the size of the population eligible and applying for benefits as well as other essential data including the socio-economic and demographic characteristics and distribution of the eligible population.*

SSA's information needs for policy decisions, therefore, suggest a need for a two tiered measurement program: (1) periodic rich and deep national data on the size, distribution, and characteristics of the working population with disabilities that permit analysis and simulation of alternative decision criteria, and (2) macro level ongoing national data to monitor the size and characteristics of the population eligible for benefits.

Maintaining periodicity of data collection is a common problem in constructing an indicator series. A government agency needs very detailed data to help administer a program addressing a social phenomenon. These detailed data, yielding themselves to years of alternative analyses, form the basis of long-range policy guidance. Large studies providing these data are supplemented by ongoing data for monitoring a small set of key indicators. One example of such a system is the quinquennial economic censuses, as benchmarks of the size and complexity of the U.S. retail sector, supplemented by periodic monthly data on retail sales, plant and equipment investments, and other variables. Another approach would be to undertake followback surveys of panels of the large survey.

Only with continuous data collection will analysts and policymakers have the information to understand and predict the impact of changes in the environment on an individual's propensity to apply for benefits and other similar issues. *SSA should make the investment in resources to expand its infrastructure to develop a permanent information-gathering system to monitor the disability-related needs of those it serves and the impact of disability benefits programs it is required to maintain.*

Recommendation 5-1: The committee recommends that the Social Security Administration develop an ongoing disability monitoring system from the experience with the National Study of Health and Activity. The committee further recommends that SSA establish a clear set of objectives for guidance in developing and implementing the substantive content of the system.

Specific objectives might include all or some of the following:

- develop the capacity to estimate the current, and project the future, prevalence of work disability and the characteristics of the population with disabilities on an ongoing basis;
- assess how well its programs are serving persons with disabilities;
- monitor the number and proportion of working age adults with impairments severe enough to apply for benefits;
- monitor allowance rates at all levels of adjudication and investigate reasons for variation across regions and over time;
- monitor changes in nonmedical risk factors associated with the application for benefits, including changes in demographic characteristics, nature of employment, and nature of disability compensation programs outside of SSA; SSA should be able to observe the

impact of changes in the demand for labor, nature of work, and other risk factors on the propensity of individuals with impairments to apply for benefits; and

- be able to foresee change in demand for benefits by identifying the precursors to change, such as the nature of employment, personal and public attitudes about coping with disability, and alternatives to SSA disability benefits, so that it is better able to anticipate the need for its disability benefits.

Underlying these objectives, the committee believes that SSA must be aware at all times of the potential need for, and effectiveness of, its disability benefits. It must know how many Americans may need their benefits, who has been applying for them and why, how satisfied those receiving them are with the administrative apparatus that has been installed to deliver benefits, and why those eligible but not receiving benefits have not applied.

A disability monitoring system would begin with a comprehensive measurement such as the NSHA, from which a reduced set of indicators of the size and characteristics of the "pool" of applicants for disability benefits would be identified. Every *n* years a similarly large and in-depth survey would be mounted. In the intervening years the reduced set of indicators (or estimations based on existing data) would be the source of national estimates of the size and distribution of the potentially eligible persons and of other issues. With each passing year the relevance of the comprehensive survey's data and analysis declines. The magnitude of policy and social changes in the intervening years affects the periodicity of the comprehensive survey. If policy or social changes are large, SSA may need to mount another comprehensive survey of the richness provided by NSHA within a three- to four-year period. If changes are small, the periodicity of the comprehensive study might be extended to, for example, every 7 to 10 years. This periodicity also allows sufficient time for thorough evaluation, planning, and testing innovations. The next iteration of the NSHA-type survey with rich measurements conceivably might use a new set of measures, developed as a result of scientific progress or changes in program direction since the last large disability survey.

Recommendation 5-2: The committee recommends that the disability monitoring system consist of

- **a periodic, comprehensive, and in-depth survey to measure work disability; and**
- **a small set of core measures in the intervening years derived from other surveys, reinterviews, and/or administrative data.**

SSA should collaborate with other federal agencies on the design and implementation of the monitoring system.

CHARACTERISTICS OF A DISABILITY MONITORING SYSTEM

The committee defines a disability monitoring system as an ongoing systematic collection, analysis, and interpretation of data essential to the planning, implementation, and evaluation of the Social Security Disability Insurance (SSDI) and the Supplemental Security Income (SSI) programs, closely integrated with the timely dissemination of these data to those who need to know.

Monitoring systems typically rely on a variety of data sources originally designed for other purposes such as, but not limited to, the national surveys of the National Center for Health Statistics (NCHS), the Bureau of the Census surveys, and administrative data. No single standard exists in the design of monitoring systems; rather, they should be designed to meet the specific purposes of the specific system. Developing a monitoring system is not dissimilar to the design of a complex survey consisting of multiple components. The components depend on the objectives of the system. The utility of a monitoring system is a function of the extent to which the data are used to make decisions, set policy, or implement changes and is evaluated in terms of the objectives of the system.

Design of a Monitoring System[1]

Disability (for adults) is defined in the Social Security Act as inability to engage in substantial gainful activity because of a medically determinable physical or mental impairment lasting at least 12 months. Therefore, one of the challenges in thinking about the design of a disability monitoring system is to understand how shifts in the nature of work over the past 40 years and into the future affect the meaning of disability, and to make this operational in household surveys and administrative databases. As stated earlier, of interest in the disability monitoring system is not simply the measurement of prevalence and the socioeconomic and demographic conditions linked to disability, but also understanding changes in both the individual and the environmental factors that lead to changes in appli-

---

[1]Much of the information in this section is drawn from the background paper "Survey Design Options for the Measurement of Work Disabilities," commissioned from Nancy Mathiowetz for use by the committee. The committee appreciates her contribution. The full text of her paper can be found in Part II of this report.

cations. For instance, the system will need to estimate the prevalence of persons eligible for disability benefits as well as develop measures that predict application. Key to such a system will be sufficient data to understand macro- and micro-level factors that distinguish participating and nonparticipating eligible populations.

The design of a disability monitoring system must consider the informational needs of the system and the impact of alternative design options on meeting analytic goals as well as the impact on various sources of survey error (e.g., whether the design should include the use of household surveys). Alternative design components include the following:

- *Data source or sources:* Among the various data sources that could be included, alone or in combination, in the design of a disability monitoring system are data obtained from household-based surveys, physical examination, and administrative records. Among the options with respect to household data are stand-alone surveys that permit rich and deep national data on the size of the disabled population (e.g., similar to the NSHA), survey modules administered as part of preexisting data collection efforts (e.g., a supplement to the Current Population Survey or the Survey of Income and Program Participation), or the incorporation of a limited number of questions on existing national surveys (e.g., the National Health Interview Survey or the Behavioral Risk Factor Surveillance System [BRFSS]). Each of these options has implications for the error properties of the resulting estimates, including coverage, sampling, nonresponse, and measurement error. The use of administrative record data potentially suffers from similar sources of error.

- *Periodicity of measurement:* Decision on periodicity requires answers to several questions, such as: If survey data are collected, how often should the data collection occur? What are the ramifications of more frequent or less frequent data collection on the utility of the data? How is periodicity affected if one decides to use repeated cross-sectional data collection or a longitudinal design?

- *Mode of data collection:* For survey data collection, a decision will need to be made as to the mode or modes of data collection—such as telephone or personal interviews, self-reports, or observation and examination. Little is known about the effect of mode of data collection on the measurement error properties of self-reports of disability and impairments. In addition, the choice of a single mode of data collection has potential implications for the coverage of the population and the potential for nonresponse bias.

- *Self and proxy response status:* Questions that need to be resolved include: Should only self-response be accepted for household surveys related to disability? If so, what are the ramifications on nonresponse bias? If proxy responses are accepted, what impact does this design choice have on the measurement error properties of the reporting of disability?

As is evident from the alternative design components discussed above, each choice impacts the error structure of the estimates of disability and the analytic capabilities that can be addressed with the resulting data. SSA will need to decide how much error both ways it will be willing to tolerate, taking into consideration costs, information needs, and other factors. Also evident is the lack of information with respect to the specific impacts of design choices on the reporting of impairments and disabilities.

One could consider a number of various permutations of design choices outlined above in designing a work disability monitoring system. These options could be arrayed along lines of richness of the data, quality of the data, and costs. At one end of the spectrum is a monitoring system characterized as a continuous, longitudinal, multimode household-based data collection, which may be supplemented periodically with medical examinations (for those meeting a particular threshold based on the household data and a subset of those who are classified in the category adjacent to the threshold) and links to administrative records. Such a design would facilitate analysis of change over time in the size of the pool of eligible population and applicants, as well as understanding of the individual and environmental factors that influence application for ben-efits, and would simulate the impact of alternative decision processes, provided that the household survey, medical examination, and adminis-trative records collected or contained the information necessary for such modeling. This comprehensive design would be the most costly.

At the other end of the spectrum are data characterized by a small number of questions on disability included as part of repeated cross-sectional surveys. Such a design would allow analysts to monitor the size of the pool of eligible population, and possibly, if crosswalk analytic capabilities had been developed, the size of the pool of applicants. How-ever, it does not facilitate understanding of how individual, environmen-tal, and macro level changes impact the application process. This minimal design would be least costly.

Design Options for Continuous Monitoring

Continuous monitoring should be undertaken using one or more de-sign options; each of which requires some statistical coordination. These may include

- sponsoring annual surveys based on self-report data from a reduced set of disability measures;
- funding additional survey questions, suitable for estimation of the size of the population eligible for disability benefits, as part of, or supplement to, an ongoing household survey;
- longitudinal data collection; forming a partnership with other on-going surveys;
- linking survey information with administrative databases; and
- ad hoc special studies.

1. Reporting from Reduced Sets of Measures in the Intervening Years

The NSHA should yield a complex and large set of measures that are used to identify alternative estimates of the number, distribution, and characteristics of the working age population in the United States potentially eligible for benefits under the Social Security disability programs. It is likely that the set of NSHA variables used to compute the "best" estimate of the pool of the eligible population would be too large to be feasible in ongoing monitoring because of the time needed for, and the high cost of, mounting a survey with such measures frequently.

How large a set is needed to attain stable estimates? Sensitivity and specificity criteria often favor different subsets of indicators. In any case, the practical problem for SSA is the issue of how large a data collection budget can be allocated to ongoing measurement of these indicators. One key principle of an ongoing monitoring system for disability is the cost efficiency of measuring a small number of attributes continuously. These could probably be self-report measures that require only a few minutes of interview time for the respondent. Thus, the ongoing measurement will be less expensive to support than the large, comprehensive periodic disability surveys.

What indicators should be measured continuously, and what should be measured less frequently? The set of measures in the periodic surveys defines the population of items from which the smaller set of continuous measures would be identified. Statistical analysis of the "best" sets of variables can be conducted (using item response theory notions or more traditional predictive analysis) with the goal of identifying a smaller set of measures that might be used more routinely to estimate the size of the eligible pool. Conceptually the problem of identifying the best subset of indicators devolves to measuring what portion of true eligibles is identified as *eligible* by the reduced set of measures (sensitivity) and what portion of actual ineligibles is identified as *ineligible* by the reduced set (specificity). The success of the ongoing monitoring measures depends on the success of the large periodic surveys.

After a reduced set of measures from the large periodic survey of a size manageable by SSA is identified for ongoing monitoring, two additional methodological steps are required. First, simulation of sample design requirements must be conducted for minimal levels of sampling variance of ongoing estimates of the size of the pool. Such simulations should provide the effective sample sizes required for all subpopulations of policy relevance, in order to inform the policy and budget functions of SSA. Some of the issues for consideration in the sample design simulations will be the needed frequency of national estimates and the desired sensitivity of the monitoring efforts to changes over time in the size of the pool of the eligible population. For example: Is it necessary to know the size of the eligible pool at any one point in time within a 1 percent tolerance, a 10 percent tolerance? If there is a 5 percent change in the size of the pool across adjacent years, must SSA be able to detect this for program management purposes? Are separate estimates required of the size of the pool for different age groups, regions of the country, occupational subgroups, or gender?

The second methodological step required after the reduced set of measures is identified, is a test of the measurement performance of the reduced set. It is common to find that 10 items extracted from a set of 100 perform differently by themselves in survey measurement than they did in the context of the 100 items. That is, the size of the "pool" estimated by NSHA using the reduced set of items might be somewhat different from that obtained when those items are introduced into another survey context. To determine the sensitivity and specificity of the reduced set of items, SSA will need to test them on a set of respondents whose eligibility is or can be known. Such studies are expensive and are often restricted to small samples, with high internal validity at the cost of low external validity. Once the reduced set of items proves its worth on its own, the set of items is ready for production use.

To summarize, SSA will need to (1) identify, through analysis of NSHA data, the set of NSHA variables needed to provide the best prediction of eligibility for NSHA respondents; (2) estimate the size and design of samples that would achieve desired levels of precision for the estimates of the pool; (3) test the reduced set of measures in a design permitting estimation of sensitivity and specificity; and (4) determine the SSA budget that can be allocated for ongoing monitoring of the size of the eligible pool. Once these tasks have been completed, SSA can then examine alternative ways to mount an ongoing monitoring of the size, distribution, and characteristics of the population eligible for benefits.

There are two common ways in which program agencies monitor the size of the potential pool of program participants: (1) the agency sponsors ongoing surveys to estimate the pool; and (2) the agency enters into a partnership with another survey to add a small set of measures in return for financial support of that survey.

2. SSA Sponsoring Surveys of Disability

Under this option, SSA would design an ongoing survey whose goal would be the estimation of prevalence of benefit eligibility. To be useful it would include as measures, those questions found most predictive of eligibility in the NSHA.

A decision by SSA to sponsor an independent ongoing survey to estimate the size and characteristics of the eligible population would be based on answers to several questions such as: What is the scope of SSA's ongoing information needs that can be combined with those of disability monitoring, and what are the administrative, financial, and technical staffing burdens of designing and estimating operation for an ongoing survey? What are the results of the statistical analysis to identify a subset of measures and how detailed a measurement is required to achieve minimally acceptable sensitivity and specificity parameters? Is a separate ongoing survey needed or can the needed subset of measures be obtained from an existing continuing survey? If there are many characteristics of the population that are not now being well described in the existing surveys, then a separate SSA survey may be justified as a small fraction of the total funds allocated to fulfill its mission.

3. SSA Forming a Partnership with Other Ongoing Surveys

An alternative to sponsoring an ongoing survey is to add a limited set of work disability indicators to an ongoing survey sponsored by another agency. This alternative would provide continuous information about the size of the eligible population by forming a partnership with a household survey of sufficient periodicity and size. In this option, SSA would, through an interagency transfer of funds, support the testing and implementation of a short set of questions that would provide prevalence estimates as add-on or as supplement to the regular survey. This option differs from those above in that it offers SSA some control over the data used in the monitoring effort but less control in terms of content, definitions, and timing than offered by its own ongoing survey.

Survey Partners in Disability Monitoring. Several federal statistical agencies currently include some measurement of disability in one or more of their household data collection efforts; several other statistical agencies are currently developing such measures for inclusion in their studies. The candidate surveys for ongoing monitoring include the American Com-

munity Survey, the American Housing Survey, the Current Population
Survey, the Medical Expenditure Panel Survey, the National Crime Victimization Survey, the National Health Interview Survey, the National
Health and Nutrition Examination Survey, the National Household Survey of Drug Abuse, and the Survey of Income and Program Participation.
Table 5-1 presents a summary of the attributes of the candidate surveys
listed above. (More detailed descriptions of these surveys and their attributes can be found in Mathiowetz, 2001, in Part II.)

TABLE 5-1 Federal Data Collection Efforts

Survey	Sponsor/ Contractor	Topic	Sample Design
American Community Survey	Census	Demographics, housing, social and economic characteristics	Rolling sample of addresses
American Housing Survey	Census	Housing, household characteristics, income, recent movers	Fixed sample of addresses selected in 1985, plus new housing units
Behavioral Risk Factor Surveillance System	CDC/state health departments and contractors in U.S. states and territories	Preventive health factors and risk behaviors linked to chronic diseases, injuries	Varies by state; probability samples of households with telephones
Current Population Survey	BLS/Census	Labor force participation, employment; supplements on various topics	Rotating panel of addresses
Medical Expenditure Panel Survey	AHRQ/Westat	Health care utilization, expenditures, health insurance coverage	Continuous, overlapping panels
National Crime Victimization Survey	BJS/Census	Criminal victimization	Panel of addresses

Three criteria are used for selection of the candidate surveys discussed here: (1) each represents an ongoing federal data collection effort; (2) the sample size is sufficient, on an annual basis, to support SSA data requirements; and (3) the survey instrument currently includes, or is planned to include, measures of disability as part of the questionnaire. Some candidate surveys do not meet all three criteria but are included for consideration because of some unique design feature of the study. For example, the annual samples for the National Health and Nutrition Exami-

Sample Size	Frequency	Mode
3 million households annually	Monthly	Self-administered; mail delivery
National sample: 55,000 housing units	National: semiannual	Face to face
Metro sample: 230,000 housing units	Metro: 1/4 of the sample each year	
Adults, ages 18 and older Sample size varies by state by year	Monthly data collection; annual estimation	Telephone
59,000 households; 94,000 persons ages 16 and older	Monthly	Face to face and telephone
6,000 households, 15,000 persons per panel; panels can be pooled to produce calendar-year estimates based on 12,000 households and 30,000–35,000 persons	Panels interviewed five times over 24 months; annual estimates	Face to face; telephone
50,000 households; 100,000 persons ages 12 and older	Biannual; annual estimation	Face to face; telephone

continued

TABLE 5-1 Continued

Survey	Sponsor/ Contractor	Topic	Sample Design
National Health Interview Survey	NCHS/Census	Health care utilization, conditions, health behavior; adult- and child-specific questionnaires	Repeated cross section of addresses
National Health and Nutrition Examination Survey	NCHS/Westat	Health status, including medical examinations	Annual cross-sectional samples
National Household Survey of Drug Abuse	SAMHSA/RTI	Drug and alcohol use	State-level cross-sectional samples
Survey of Income and Program Participation	Census	Program participation and eligibility, income; topical modules by wave of interviewing	Panel of households

NOTE: AHRQ = Agency for Healthcare Research and Quality, BJS = Bureau of Justice Statistics, BLS = Bureau of Labor Statistics, CDC = Centers for Disease Control and Prevention, NCHS = National Center for Health Statistics, RTI = Research Triangle Institute, SAMHSA = Substance Abuse and Mental Health Services Administration.

SOURCE: Adapted from paper commissioned from Nancy Mathiowetz, see Part II.

nation Survey (NHANES) and the Medical Expenditure Panel Survey (MEPS) are relatively small as compared to some other surveys ($n = 5,000$ and $n = 15,000$ persons annually, respectively); however, each of their designs benefits from a complementary component. In the case of the NHANES, the design includes a medical examination. In the case of the MEPS, the design includes data from medical care providers and providers of health insurance. Similarly, the National Household Survey of Drug Abuse (NHSDA)—a large survey ($n = 67,000$) producing state estimates—does not presently include any measures of functional limitation or disability; however, the design includes both an interviewer-administered questionnaire and a self-administered set of questions that may be beneficial in the assessment of disability. The National Health Interview Survey (NHIS) is a large continuing survey of approximately 43,000 households including 106,000 persons. The NHIS sample is drawn from each state but is too small to provide state-level data with acceptable precision.

Sample Size	Frequency	Mode
40,000 households annually; 98,000 persons	Weekly replicate samples; quarterly and annual estimates	Face to face
5,000 persons	Annual	Face to face; physical examinations
67,000 persons	Annual	Face to face; self-administered
36,000 households	Quarterly; annual estimates	Face to face; telephone

The relevant questions to be addressed by SSA in choosing a partner survey(s) include the following:

1. How large a sample is interviewed each year? What standard errors are likely to be obtained for key disability prevalence statistics?
2. Will the addition of disability measures in the interview be consistent with the measurement goals of the original survey? Are there possibilities of context effects that could damage the accuracy of prevalence estimates?
3. Are there existing measures in the survey that might be used as explanatory variables for disability status indicators? Can the survey offer SSA other informational benefits beyond being a vehicle to produce disability prevalence statistics?
4. Is the survey of high quality? What evidence is there about coverage, nonresponse, and measurement error properties of key statistics?

5. How frequently can estimates be updated? Will monthly preva-lence estimates be generated, annual estimates, et cetera?
6. Is the mode of administration of the survey compatible with the measures chosen from NSHA?
7. What restrictions, if any, will SSA have on access to the micro-data from the surveys? Can SSA analysts and independent researchers use the data for other analyses of importance to SSA, or will they be given only statistics produced from the survey data?
8. If state variations in disability applications, approvals, and denials are important factors to SSA, should existing surveys, such as the Current Population Survey (CPS), and NHSDA that produce state estimates be given preference as SSA partners?
9. Will the mission of the sponsoring agency be aided by a partner-ship with SSA in measuring disability status? With the obligation of many federal household surveys to provide indicators of dis-ability, can SSA expertise in work disability be viewed as a desir-able complement to the skills of the survey sponsor's staff skills?
10. What are the cost trade-offs for SSA in choosing a partner survey? SSA must weigh the reduced costs of partnership with another federal agency to produce the information it needs against the increased costs of mounting large periodic and interim surveys for which it would have complete control.

The ideal partner(s) survey would have a sufficiently large sample to provide SSA with prevalence estimates that were stable enough to protect policymakers from erroneous results. It would have very low coverage and nonresponse errors. It would be conducted frequently, giving SSA the ability to model seasonal effects in the size of the eligible pool and to estimate the impact of economic shocks. It would contain other measures that would be of utility to SSA in addressing other important manage-ment problems (e.g., Are all demographic subgroups changing the dis-ability prevalence in the same way over time? What are the major health and demographic correlates of disability status?).

The chief obstacle to the feasibility of this partnering option for ongo-ing monitoring is that most federal household surveys are already using long and complex instruments, filled with measures of great value to existing constituencies. Seeking to add measures to these instruments faces zero-sum conflicts with existing obligations of the sponsors. The single most important sign of optimism is that several of the surveys are facing mandates to begin measurement of disability status in order to learn how the disabled subpopulation differs from others on the key topics covered by the surveys.

4. Longitudinal Data Collection

A longitudinal design, either independently sponsored or in collaboration with one or more other federal agencies, offers analytic capabilities that are not possible with repeated cross-sectional designs. This is especially the case for those designs related to the decision to apply for benefits, including both individual factors that influence the decision and the impact of environmental and macro level changes (e.g., economic) on the decision to apply for benefits.

Selected reinterviews from large intermittent national surveys could provide needed information to assess change in status in different age, occupation, and gender groups. Much more can be learned from studying changes in individuals and their environments than from one-time cross-sectional measurement research. Such a design has high response rates, more ease in locating, and often better response reliability. Particularly where expensive screening was required for the initial sample, it need not be repeated and further subselection at different rates is possible. For example, one might follow all those currently on disability rolls, half of those with disabilities but not covered, a small fraction of those not reporting disabilities but with some health problems, and a still smaller fraction of the remainder of the population. What this means is that a combination of periodic large national survey with screening, and efficiently designed follow-up mostly by telephone, could continue the research on the disability policy questions, and the effectiveness of the process for determining eligibility for disability benefits.

The committee also suggests that SSA consider sample cohort rotation and integration with other federal surveys for the design of its disability monitoring system. Since samples with planned overlap over time perform more effectively in measuring change than independently drawn samples at each time point, some sort of cohort feature might be considered for the system. Several possibilities in decreasing order of statistical effectiveness are cohorts at the person, address, and cluster levels. A cohort in which the same persons are followed over time has the advantage of following those for whom disability is measured although the cost of follow-up can be extremely high to retain a high percentage over time. Drawing respondents from the same first-stage sample cluster (or primary sampling units) is the least costly of these options but also the least advantageous statistically since clusters often account for a relatively small part of the total variation in disability measures. A compromise to these two extremes is to return to the same sample of addresses for each round of a continuous sampling process. This approach is operationally effective since one returns to the same place each round (although address samples must be updated to accommodate new construction), but people move, so only a portion of the sample is retained from one time to the next.

Longitudinal designs require that additional decisions be made concerning the length of the panel (that is, the number of years individuals are followed) and the frequency of data collection. In addition to the questions outlined above describing a periodic rich data collection effort supplemented by monitoring of the population through an abbreviated set of measures, the development of an ongoing, continuous panel design would have to address: the size of the sample needed to achieve analytic capabilities for a single calendar year versus pooled estimation across contingent years; the life of a single panel, that is, the number of years individuals will be followed through time; the periodicity of the data collection; the acceptability of mixed modes for data collection and its effect on the measurement and nonresponse properties of the resulting estimates; the use of a panel design requiring consistent response from the same respondents versus a mix of self and proxy response over time; and the ramifications of the decision on the error properties of the estimators. Several panel designs among the federal data collection efforts are shown in Table 5-1.

However, pure person and cohort samples also have the disadvantage of higher respondent burden since respondents will be asked to participate in several rounds of data gathering. To control the added burden of fully retained cohort samples, some type of rotating cohort sample might be used in the design of a disability monitoring system. For example, SSA might consider something comparable to the 4–8–4 rotation scheme that has been used by the CPS, in which a sample household is sampled for four consecutive months, not interviewed for eight months, then interviewed again for four consecutive months (Census and BLS, 2000). A very different design—a continuous overlapping panel design—is used in the Medical Expenditure Panel Survey. In that survey, members of a panel are interviewed five times over 24 months; a new panel begins at the start of each calendar year so that panels can be pooled to produce calendar-year estimates.

5. Linking with Administrative Files

Survey data can be made richer by linking with appropriate administrative files maintained by SSA for both the SSDI and the SSI programs. Administrative data usually have no information on persons who have not applied for benefits and little information on socioeconomic variables. Household population surveys, on the other hand, provide information on persons who have not applied and on a wide range of socioeconomic variables but contain little or no information on the person's interactions with the administering agency (Hu et al., 1997). Some examples of the administrative files that can be used are

- The *Master Beneficiary Record* (MBR) is the main file administrating Social Security retirement and disability insurance payments. It contains the data to administer the SSDI benefit program. The MBR record is initiated once the initial decision is made to award benefits, and entitlement and payment data are stored in the file.
- The *Supplemental Security Record* (SSR) is the main file for administering the SSI program. It provides the data needed to generate federally administered SSI benefit checks. SSA establishes and updates the SSR through local field office and teleservice site transactions, usually establishing the record as soon as a person files for SSI. The file stores eligibility and payment information.
- The *Master Earnings File* (MEF) contains earnings records for calendar years since 1951 and contains approximately 400 million records. Since 1977, the MEF has been derived primarily from Internal Revenue Service (IRS) Form W-2 data. MEF data are used for computing SSDI benefits. In addition to the MEF, the SSR is a source of earnings data from SSI applications.
- The *831 Disability file* is related to both SSDI and SSI. When a person applies for disability benefits from either program, a medical determination is required. Medical decisions are made by the Disability Determination Services, reported to SSA's Office of Disability (OD), and recorded in the National Disability Determination Services System. The 831 Disability file is extracted regularly from this system for research purposes. The 831 Disability file may contain data from decisions made at several levels of adjudication that represent ever-higher levels of appeal. Most records in the 831 files pertain to only two levels: (1) the initial medical determination; and (2) the reconsideration decision (i.e., the first level of appeal for medical denials). Decisions made as a result of SSA's Continuing Disability Review process are recorded on OD's 832 and 833 Disability files. The unit of observation in the 831 Disability file is a disability decision; the main data elements capture the primary impairment code, the regulation basis code (used to measure the severity of the impairment), the date of decision, the level of decision, and the result of the decision.
- In addition to the master files, the *Continuous Work History Sample* (CWHS) file is a 1 percent continuous work history sample from 1937—when payroll tax was first levied—until the present. The intent of the sample is to measure the working trends and employment of the population in relation to the Old-Age, Survivors, and

Disability Insurance (OASDI) program. The data in the CWHS are drawn from other master files—the Numident, MBR, and MEF—and from the IRS self-employed file. The CWHS also contains derived and constructed data elements that do not appear on any master file. Given the stringent restrictions on direct MEF access, the CWHS has become a substitute for the MEF in many instances. As of June 2000, access to the CWHS is very limited, but efforts are under way to make the file more widely available.

6. Ad Hoc Special Studies

In its plans for a monitoring system SSA should include ad hoc studies on specific emerging policy issues as well as explore other questions that do not need continuing data. One example is a follow-up study of applicants for disability benefits to see whether some years later they are working (disabled or not), or not working (even if they were denied benefits). The ratio of denied who remain not working to the accepted who could be, or are, working, is some indication of the accuracy of the decisions. The total number of errors both ways could be some indication of the efficiency of the system. For example, Bound (1989) using data drawn from the 1972 and 1978 surveys of the disabled done by SSA, found that fewer than 50 percent of rejected male applicants work.

Another example is a study of employers as well as people with disabilities to develop information on employer tolerance in hiring workers with disabilities and on the willingness of employers to display the flexibility often required to deal with workers with disabilities. SSA should explore the experience of other agencies in conducting such surveys. The National Center for Health Statistics had conducted a survey of employers and the Medical Expenditures Panel Survey also surveys employers.

DEVELOPING AND PLANNING A WORK DISABILITY MONITORING SYSTEM

The workshops held by the committee and input from experts in the field led to a clear conclusion that key concepts in disability were subject to debate among scientists, policymakers, and disability interest groups. Comparisons of U.S. social legislation on disability and that of other nations, arguments about the role of the social and physical environment of a person in defining disability, and the impact of macroeconomic forces on self-identification as work disabled, all led the committee to concerns that the concepts and measurements of disability over time could (and perhaps should) undergo change.

The prospect that key constructs will force new measurement challenges over time is a common problem in social measurement. The chief model to address this problem in other fields is to form partnerships with the scientific field allied to the constructs being measured. An ongoing program of measurement development is needed, allied with the conceptual developments in the field. Such research, when conducted outside an agency but guided by the mission of the agency, can supply the agency with proven measurement approaches when the new concepts become integrated into statutes guiding program designs. For example, small-scale studies examining how environmental impacts on disability self-reports manifest themselves can be valuable to the development of structured survey questions. The Disability Research Institute is one possible locus for such research.

Because notions of disability and models of influences on disability are constantly changing, any ongoing monitoring system to monitor the phenomena must adapt and change over time. This can be accomplished only with ongoing investment in new methods of measurement.

Planning for a Monitoring System

In order to develop a monitoring system in collaboration with other relevant agencies, the following elements are necessary:

- SSA should set aside a multiyear planning period to systematically design and establish the proposed disability monitoring system;
- the system should be designed with sufficient flexibility to accommodate the evolving medical, legal, social, and policy perspectives of disability;
- the system should use as much as possible the design and data from existing federal surveys that measure disability in the population (e.g., NHIS, NHANES, the American Community Survey), by further cultivating partnerships with the agencies that conduct these surveys; and
- SSA should ensure the availability of sufficient qualified research staff to design and oversee the proposed disability monitoring system.

While data gathering and analysis of the NSHA are under way, the committee encourages SSA to begin planning a national disability monitoring system to serve as its main information source for program planning and assessment. The general goal of the monitoring system would be to continuously monitor the number of Americans who are eligible to receive SSA disability benefits (i.e., the size and characteristics of the population eligible for disability benefits) as well as public access to, utilization of, and satisfaction with the nation's disability benefits programs. Although resources may dictate many of the system's design features, the committee urges SSA to make its system one in which data are collected on a continuing basis from a valid and statistically efficient national sample of households.

SSA should consider using data from existing federal surveys in designing the monitoring system and should use these information sources to supplement data generated from the SSA system. Besides serving as a supplementary source of disability data, integration with the design features of these surveys might prove beneficial both to SSA's monitoring system and to the other surveys. This level of design integration has already been successfully accomplished between the Medical Expenditure Panel Survey conducted by the Agency for Healthcare Research and Quality and the National Health Interview Survey conducted by the National Center for Health Statistics (Cohen, 1999).

To design and establish a high-quality SSA disability monitoring system, the committee suggests that a planning period of approximately three years would be needed. Three important tasks must be completed during this planning phase. First, SSA must recruit a sufficient cadre of qualified staff to conduct the design work and directly oversee the initial field-testing of the system largely from within SSA. The implication is that SSA should develop this system largely from within and not rely heavily on external contractors to do the work. Among those professional expertise areas that SSA would need to recruit are experts in: functional ability, cognitive measurement, survey design, and analysis.

The next planning phase would be for SSA to develop a detailed blueprint for the system. This step suggests that pilot testing an early version first in a few states or metropolitan areas before national implementation might best phase in the system. The goal here would be to produce a system design that is useful to SSA, scientifically sound, and able to withstand careful scrutiny, yet sufficiently flexible to adapt to changing information needs over time. Toward these ends, the committee suggests that SSA support a careful study of the cognitive and process effects of measuring disability in a survey context. At a minimum this research should answer the following questions: What are the effects of the mode of gathering the data (e.g., self-administered, telephone interview, face-to-face interview)? How is the portrayal of disability influenced by who provides the data (e.g., the disabled person, a proxy caregiver, a health care provider)? How do question wording, context, and format affect the picture of disability painted for a disabled person? Ultimately, the role of this research on measurement effects is to understand the effect of the survey design strategies used to measure disability in NSHA and other major surveys and thus enable SSA to interpret disability data from all current sources.

Suggested steps in a multiyear planning and implementation schedule starting in 2003 shown earlier illustrate a strategy that is divided into three, sometimes overlapping phases involving strengthening the infrastructure and developing interagency collaboration, initiation of research on design, and a testing phase. These steps are shown below in Box 5-1.

The committee recognizes that despite its many benefits, developing and implementing the recommended disability monitoring system raises several important issues that would require careful examination and resolution during a three-year planning period before final decisions can be made on the details of the design. Many of these issues relate to conceptual definition, method, timing, collaboration with partner agencies, and resource requirements.

The committee has provided SSA with a conceptual framework with alternative choices for SSA to decide on; it has also recommended that further research be undertaken on unresolved methodological and logistical issues to reach informed decisions on implementing the details of the design. In the committee's opinion, SSA would benefit from technical guidance provided by an external group of technical experts.

BOX 5-1 Suggested Implementation Schedule

Phase I: 2003–2004
- Explore agency collaboration
- Obtain necessary funding for extramural research
- Ensure a cadre of qualified research staff
- Select expert technical advisory committee

Phase II: 2004–2006
- Develop operational definition(s) of work disability
- Conduct studies relating to design features
- Search for and test performance of "best" set of measures
- Investigate statistical benefits of overlapping designs
- Develop model-based respondent imputation strategy
- Decide on design options

Phase III: 2006
- Design and pilot-test interim monitoring component
- Analyze results and make needed adjustments

Recommendation 5-3: The committee recommends that the Social Security Administration establish a continuing, external technical committee of experts for the planning and implementation of the recommended disability monitoring system.

One model for such a committee is the working group established under the auspices of the Section on Survey Research Methods of the American Statistical Association, which currently advises the Centers for Disease Control and Prevention (CDC) on the BRFSS and the Census on the Survey of Income and Program Participation. This arrangement has been an effective vehicle for the CDC and Census; it could serve as a model for SSA to consider in carrying out the above recommendations.

In conclusion, the committee emphasizes that developing and implementing a work disability monitoring system as recommended in this chapter will contribute toward a significantly improved and efficient system of measuring and monitoring work disability and effective fiscal management of the programs.

6

Improving the Disability Decision Process

This chapter briefly discusses the key issues identified in the committee's preliminary assessment of the Social Security Administration's (SSA's) research plan to redesign the disability decision process (IOM, 1998), and the subsequent decision by SSA to terminate this redesign effort and explore ways to incrementally improve the current process. The chapter then discusses and makes recommendations on the research needed to bring about fundamental improvements in the current disability decision process.

BACKGROUND

Determination of eligibility for disability benefits under the Social Security Disability Insurance (SSDI) and Supplemental Security Income (SSI) programs is an inherently difficult task. To qualify for benefits under these programs a person must have a medically determinable impairment. Although the existence of a medically determinable impairment is a necessary condition, it is not a sufficient condition for receipt of benefits. The statutory definition makes clear that this program deals with *work disability*. The applicant is considered to be "disabled" (as defined by the Social Security Act) not just because of the existence of a medical impairment, but because the impairment precludes gainful work (Hu et al., 1997) anywhere in the national economy, taking into consideration the person's age, education, and work experience, which are commonly referred to as vocational factors. Disability determination is a complex process, inescapably involving some interpretive judgment about capacity for work (GAO, 1994; Hu et al., 1997). At a minimum, making such decisions requires clinical determination of the extent of a claimant's physical, mental, or sensory impairments; analysis of the degree to which such impairments limit the claimant's functional capacity relevant to work roles; and consideration of the interaction of the claimant's physical, mental, or sensory impairments with the person's age, education, and work experience to provide an overall picture of the claimant's future capacity for any sort of work. Finally the disability decision process requires a means for comparing those capacities with the capacities demanded by work roles in all jobs in the national economy that provide substantial gainful activity (SGA) earnings level.

While many of the factual determinations are relatively straightforward, others range from the difficult to the nearly impossible. For instance, while measures of visual acuity are reasonably well understood and can be readily translated into sensory limitations, the measurement of pain and its effect on function is much less amenable to objective determination. The real demands of jobs in the national economy are constantly shifting in ways that make straightforward measures of functional capacity problematic guides to a worker's true capacity for success in the workplace. Therefore, it is impossible to know precisely the extent of imperfection in the determination of disability, as evidenced by the lack of agreement observed in an examination of rater reliability as measured by the variations within and between states in the allowance rates by examiners (Gallicchio and Bye, 1980; DHHS, 1982). SSA has been struggling with these issues for more than 40 years in the face of high volumes of claims for adjudication (millions of claims per year decided by more than 10,000 adjudicators at various levels of the process) and high levels of legal challenge and political oversight.

THE CURRENT DECISION PROCESS FOR INITIAL CLAIMS

The standards for evaluating disability claims are specified in SSA's implementing regulations (20 Code of Federal Regulation, parts 404 and 416, subparts P and I) and in written guidelines. These regulations and guidelines describe a sequential process for determining whether or not a claimant meets the statutory definition of disability.

The purpose of developing the sequential decision process is to provide an operationally efficient definition of disability with a degree of objectivity that can be replicated with uniformity throughout the country. SSA's overall objective is to adjudicate claims as consistently, expeditiously, and cost-effectively as possible. As described briefly in Chapter 1, the disability decision process for initial claims involves five sequential decision steps (SSA, 1994a).

1. In the first step, the SSA field office reviews the application and screens out claimants who are engaged in substantial gainful activity.
2. If the claimant is not engaged in SGA, step two determines whether the claimant has a medically determinable severe physical or mental impairment.
3. During the third decision step, the documented medical evidence is assessed against the medical criteria to determine whether the impairment meets or equals the degree of severity specified in SSA's Listings of Impairments (Listings). A claimant whose impairment(s) meets or equals those found in the Listings is allowed benefits at this stage. The Listings serve the purpose of allowing rapid payment of benefits to claimants whose presumed residual functional capacity (RFC), given the severity of their impairments, would preclude work at virtually any job.
4. In the fourth decision step, claimants who have impairments that are severe, but not severe enough to meet or equal those in the Listings, are evaluated to determine if the person has residual functional capacity to perform past relevant work. Assessment of the RFC requires consideration of both exertional and nonexertional impairments. If a claimant is determined to be capable of performing past relevant work, the claim is denied.
5. The fifth and final decision step considers the claimant's RFC in conjunction with his or her age, education, and work experience to determine whether the person can perform other work that exists in significant numbers in the national economy.

The determination in the fifth step is based on the 1978 Rules and Regulations, Medical–Vocational Guidelines (referred to as the *vocational grid*). The vocational grid, like the Listings, is intended to lend objectivity to the determination process and facilitate uniform administration of the vocational portion of the disability determination process. The grid reflects only physical (exertional) impairments. It does not consider nonexertional (e.g., mental or cognitive) impairments. The regulations also recognize that some claimants will have multiple impairments or environmental limitations (e.g., they cannot be around fumes) that are not effectively covered by the grid regulations. These cases must be decided outside the grid.

Proposed Redesigned Decision Process

SSA has stated that the redesigned disability decision process should

- be simple to administer;
- facilitate consistent application of rules at each decision level;
- provide accurate and timely decisions; and
- be perceived by the public as straightforward, understandable, and fair.

The goal of the new decision process was to focus decision making on the functional consequences of an individual's medically determinable impairment(s) (SSA, 1994a). Although the presence of a medically determinable impairment would remain the central requirement for eligibility as required by law, the redesigned process would focus directly, rather than indirectly, on the applicant's functional ability to work and would rely on standardized instruments for measuring functional capacity to reach decisions. Medical and technological advances and societal perceptions about the work capacity of a person with disabilities appear to support a shift in emphasis from the current focus on disease conditions and medical impairments to that of functional inability. For example, people with disabilities are able to function today with personal assistants and assistive devices.

SSA assumed that under this proposed decision process, the majority of disability claims would be evaluated using a standardized approach to measuring functional ability to perform substantial gainful activity. Standardizing the approach to assessing individual functional ability would facilitate consistent decisions regardless of the professional training of the decision makers in the decision process. The new disability decision process, as envisioned by SSA, would assess a person's functional ability once, relying on objective, standardized, functional assessment instruments. SSA believed that focusing decisions on the functional consequences of a person's medical impairments would permit physicians and others who provide medical evidence, as well as decision makers, to use a consistent frame of reference for determining disability, regardless of the diagnosis and would facilitate evidence collection by reducing the need for developing extensive medical records (SSA, 1994a).

In the proposed plan, decision makers would consider whether a person has a medically determinable impairment(s), but would no longer impose a threshold "severity" requirement. Instead, they would compare the individual's impairment(s) against an "Index of Severely Disabling Impairments." The Index would replace the Listings of Medical Impairments (SSA, 1994a). The Index was to have described, quickly and easily, impairments that are so severely debilitating that, when appropriately documented, they can be presumed to equal a loss of functional ability for SGA without assessing the individual's functional ability and without consideration of the person's age, education, and work experience. The medical findings in the Index were to have been as nontechnical as possible and to exclude such things as calibrations or standardization requirements for specific tests and/or detailed test results. The Index was to have been easy to understand and simple enough for laypersons to understand. SSA, therefore, believed that it would no longer need the concept of "medical equivalence" that is in the current decision process, thus eliminating one decision step in the current sequential evaluation process. If the claimant was not considered eligible for benefits based on the Index, then the person's functional ability would be measured using standardized instruments or protocols linked to clinical and laboratory findings (SSA, 1994a). The effect of the statutorily prescribed factors of age, education, and work experience would be considered when deciding the claimant's ability or inability to engage in substantial gainful activity.

The sequential process as it exists today and a new process as proposed in SSA's redesign proposal are illustrated in Figure 6-1.

SSA'S RESEARCH PLAN FOR A REDESIGNED DISABILITY DECISION PROCESS

As directed by the Commissioner, SSA officials developed a research plan in 1995 for examining the feasibility of, and developing, the various components of the redesigned disability decision process. The plan had three components:[1]

1. Information gathering (comprehensive review and analysis of existing information) on

   - functional assessment instruments;
   - occupational classification systems;
   - disability determination processes used in other disability programs in the United States and other countries; and
   - the effects of age, education, and work experience (vocational factors) of the applicant.

---

[1]SSA's research plan, along with a time line for actions and completion dates, was published in the *Federal Register* in August 1996 and an update of the plan was published in November 1997 (SSA, 1996, 1997).

Current Decision Process   **Proposed New Decision Process**

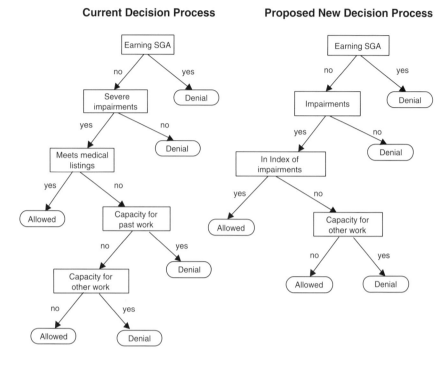

FIGURE 6-1 The Social Security Administration's current and proposed redesigned disability decision process.
SOURCE: Adapted from Hu et al., 1997, and reprinted from IOM, 1998.

2. Integration, synthesis, and development of a prototype for a new disability decision process:

   • analysis and evaluation of the literature reviews undertaken in the first component; and
   • development of prototype(s).

3. Testing, analyzing, and refining the prototype:

   • laboratory research and other small-scale testing; and
   • the National Study of Health and Activities (NSHA): a national sample survey to estimate the size and characteristics of the population eligible for disability benefits, determine factors permitting them to work, assess future changes in the prevalence of disability, and serve as one of the evaluation mechanisms for the decision process prototype(s).

Work on the first two components of the research plan was under way when this committee began its study. The research mostly involved contracts to conduct literature review and analysis in the areas of functional assessment instruments, occupational classification systems, disability determination processes used in other disability programs, and effects of vocational factors in the disability decision process. Some of the information-gathering activities were completed when the committee began its review. Work on the second component and parts of the third component involved the award of a task-order contract at the end of fiscal year 1997. The purpose of this contract was to synthesize and integrate the results of the literature review contracts and the NSHA; to develop, test, evaluate, and refine alternative prototypes for a redesigned disability decision process; and to undertake additional research as needed. Under SSA's redesign research plans, additional work in the research and development of the decision process would have been undertaken in subsequent task orders under this umbrella contract and/or under separate contracts.

In summary, SSA's research plan aimed at developing and testing the functional assessment instruments in the disability decision process, examining the effect of vocational factors on decisions, exploring what is being done in other disability programs, and developing a prototype for a revised disability decision process.

Committee's Review of the Research Plan

Early in the study, the committee conducted a preliminary review of SSA's workplan and individual research projects completed and under way. The committee also explored other relevant internal documents provided by SSA in response to requests for information. It heard presentations from the staff of SSA on work completed to date, plans to integrate the results of the research projects and the NSHA to develop a redesigned disability decision process, and the time line for completion of all research.

After reviewing the available documents and discussions with its contractors, the committee decided that a preliminary assessment of the adequacy of the research plan was needed to guide SSA management in determining whether the research activity undertaken was adequate and what more, or different, was needed to conclude whether or not the proposed revisions in the decision process were feasible, practical, and could be implemented nationally.

According to SSA, the goal of this research is to devise a more efficient and more accurate method for making timely determinations of disability for Social Security claimants (SSA, 1996). In the context of that goal, the committee outlined an initial conceptual framework of issues, research steps, and methods for a research plan or design to develop and assess the proposed new decision process as a workable solution to current problems. Such a framework should delineate at the outset the nature and extent of the problem. Complaints, whether ultimately substantiated or not, often suggest that a program should be evaluated and improved. However, in order to assess the validity of the complaints, objective evaluative criteria should be established a priori, so that the various complaints about the program can be evaluated and the program's performance can be measured. The next step would be to identify alternative solutions that might address the problems. Finally, the proposed new decision process should be tested to determine whether it is workable and whether it will alleviate the problems initially identified. To determine if a redesigned disability decision process would lead to improvements, one or more studies need to be conducted to provide information on how the current program is working relative to the established criteria. Analysis of data from such studies would identify the gaps between performance and the goals of the program. This framework is reproduced in Table 6-1.

The committee then reviewed the general features and directions specified by SSA in its research plan and the individual projects within the plan with reference to each of the research steps identified in the framework. It reviewed and commented on both the completed research projects listed earlier and the scope of work in the relevant requests for proposals for the conduct of the research. It identified critical elements of a research design that were missing from SSA's plan, expressed serious concerns about these gaps, and made recommendations for redirected and new research effort.

The committee commended SSA for initiating the major task of redesigning to improve the disability decision process and undertaking a range of research activities related to the functional consequences of medical impairments and for recognizing the need to assess the feasibility, validity, and reliability of a proposed redesigned decision process. Nevertheless, the committee concluded in its second interim report (IOM, 1998) that the research completed, under way, and planned appeared to lack the necessary overall framework and lacked the critical elements of a well-designed research plan. Some of the key conclusions reached in that report are summarized below.

In 1995, SSA contracted with the Virginia Commonwealth University (VCU) to review systems, methods, and instruments that measure a person's functional capacity to perform activities and tasks, to develop a matrix of categories to classify these instruments, and to evaluate them to determine their potential application in the disability decision process. VCU's main conclusion in its report was that no government or private entity is currently using functional assessment instruments specifically

TABLE 6-1 Issues and Methods to Be Addressed in a Framework for a Research Plan for a New Disability Decision Process

Question	Research Steps	Research Methods
1. What is the nature and extent of the problems with the disability decision process?	Needs assessment research	• Special surveys and analytic studies • Assembly of existing internal and external data • Satisfaction surveys • Analysis of data from studies using established evaluative criteria • Focus groups
2. What alternative solutions might address these problems?	Identify alternative options Small-scale testing Field evaluation	• Review and analysis of research literature • Specially targeted research • Laboratory research and pilot studies and demonstrations • Field tests • Focus groups • Process engineering assessments • National surveys
3. Will the proposed disability decision process be workable, and will it alleviate the problems?	Program evaluation and transition to implementation	• Clinical trials • Simulation • Evaluation studies of the proposed decision process using the established criteria • Cost-effectiveness studies • Tests of the new decision process in selected sites

SOURCE: Reprinted from IOM, 1998.

for determining work disability benefits, and a global measure of functional assessment does not exist that would be a valid indicator of disability for all populations currently served by SSA. Such an instrument will likely have to be developed and tested.

After reviewing the advantages and disadvantages of both functional and clinical assessment measurements for SSA's needs, VCU concluded that objective functional assessment can and should be a component of the redesigned process. VCU, however, stopped short of constructing the global measure of functional capacity; instead, it recommended several steps SSA should take in moving toward its development. The committee expressed concern that SSA had not made clear the conceptual or theoretical basis for believing that such a standardized, global instrument existed or can be constructed before launching the literature review project on the subject. Skeptical of one global, standardized, universally accepted measure, the committee recommended that SSA develop alternative research plans for development and use of functional measures in the disability decision process in the event that the proposed global standardized functional assessment instrument is not developed and tested (IOM, 1998).

In 1996, SSA contracted with the American Institutes for Research (AIR) to conduct a comprehensive review of the literature pertaining to systems and methods of classifying occupations in terms of the physical and mental capacities required, to develop a taxonomy of occupational classification systems, and to assess the applicability of systems for SSA's redesigned disability determination process. This review relates directly to one of the key elements in the proposed redesigned disability decision process, namely, assessing baseline work. The purpose of the review is to determine if a standard exists, and if not, whether it is feasible to develop one to describe basic physical and mental demands of a baseline of work. AIR concluded that while none of the occupational classification systems exactly or ideally matched SSA's needs, the Occupational Information Network (O*NET) under development was the closest match to SSA's needs. O*NET is an occupational classification system being developed by the Department of Labor (DOL) under contract with AIR to replace the Dictionary of Occupational Titles. One of the reasons AIR recommended O*NET over the other systems is because it uses level scales to measure the amount of skill needed to perform certain jobs. Incumbents choose a numeric rating based on their reading of the behavioral anchors. Cognitive and mental descriptors are also included in O*NET, but the physical ability scales that O*NET uses may not be specific enough to help SSA. Many other issues were identified by the contractor that need to be resolved before O*NET can be used for SSA's purposes. These issues are described in the committee's interim report (IOM, 1998) and are discussed further later in this chapter. The committee questioned how O*NET will be used. SSA's research design did not appear to be oriented to address this question. How does SSA plan to supplement O*NET with respect to contextual or other factors that are not well covered. There were no indications in the research plan that the gaps in O*NET would be carefully considered and specific research identified to fill those gaps. The committee also was concerned about the synchronization of timing for completion of O*NET and SSA's target completion of the research for development and implementation of the disability decision process. The committee recommended that SSA develop an interim plan for an occupational classification system in the event that the O*NET database is either not completed or is insufficient to meet the needs of a new disability decision process. It also suggested that SSA should explore entering into an interagency agreement with the Department of Labor to initiate a version of O*NET that would collect information on minimum as well as average job requirements to better serve SSA's needs to assess ability to engage in substantial gainful activity.

The committee made recommendations for changes in priorities and improvements in the research projects that were under way and others yet to be developed. *It urged SSA to adopt a rigorous research design process and to develop, early in the research, objective validation criteria and validation plans to be able to make the ultimate judgments on whether or not the proposed changes would yield the desired results. In issuing the interim report the committee hoped that the recommendations embodied in that report would be incorporated in the contract research that was under way and in new research not yet initiated at the time.* The committee's detailed discussion, findings, conclusions on the various issues, and recommendations are embodied in its second interim report to SSA (IOM, 1998). All the recommendations flowing from the committee's preliminary assessment are included in Appendix C.

"POST-REDESIGN" PERIOD

The committee had planned to examine and comment further on the adequacy of the entire research plan when completed, the results of the completed research, and any subsequently initiated research for the redesign effort. However, after the committee issued its second interim report (IOM, 1998), SSA undertook an internal reevaluation of its disability decision process redesign initiatives. SSA concurred with several of the committee's conclusions and some of the recommendations. However, rather than undertaking the additional research and redirection of the research as recommended by the committee, SSA decided to no longer actively pursue the new decision-making process proposed in Disability Redesign, but to improve the current process, focusing at this time on updating the Listings (SSA, 1999b). According to SSA, its new strategy is to concentrate on improving the overall adjudication process to ensure that decisions are made as accurately as possible, that those applicants who should receive benefits should get them as early as possible, and that the adjudication process is consistent throughout (SSA, 1999b).

To make improvements, SSA has stated that it has redirected and refocused disability policy development and related research activities in order to address both the longer-term goals of redesign, and the more immediate, pragmatic needs of the disability programs (SSA, 1999b). SSA expects these process initiatives to

- enhance the quality of decisions at all levels;
- streamline the disability process by applying the lessons of the disability process redesign efforts; and
- update medical and vocational rules used in making disability determinations.

The initiatives to achieve the first two objectives relate to components of the reengineering process that are outside the purview of this committee. The third objective—updating the medical and vocational guidelines—is a direct outgrowth of the now abandoned initiative to develop a new decision process.

Updating the Listings

SSA states that although it is no longer focusing on development of the new decision process described in the disability redesign plan, it is continuing to explore the potential in some of those ideas. However, it is now devoting most of its resources to needed improvements to the current evaluation process (SSA, 1999b).

SSA further states that the proposed new decision process was intended in part to address concerns about the current Listings by replacing it with an Index of Impairments. However, SSA has concluded that the Listings serve a vital role in ensuring the cost-effectiveness of the adjudication process. Currently about 60 percent of the allowance decisions are based on the Listings without developing and conducting a complete, in-depth functional and vocational analysis. Rather than replacing the Listings, SSA is now engaged in a concentrated effort to update and improve these medical guidelines (SSA, 1999b). Medical advances in both diagnosis and treatment have made updating the Listings long overdue. As stated by SSA, the general approach to revising a section of the Listings is to begin with its adjudicative experience and program knowledge. Having identified an area of interest, medical literature is reviewed and, as warranted, experts in the field are consulted. If more extensive research is needed, a contract(s) may be negotiated to obtain the information. Medical experts from within SSA are consulted to develop an initial proposal.

Functional Consequences of Impairments

SSA informed the committee that "the proposed new decision process was to have relied on simple, objective readily available functional tools to assess an individual's ability to work. Initial research conducted under contract by SSA has not shown a basis for believing that such a standardized, universally accepted global instrument applicable to individuals with physical and mental impairments can be constructed or that some similar approach would be possible. Further, some of the other assumptions on which the proposal for a new process was based (e.g., the assumption that these functional tools would be routinely used in clinical practice and, therefore, readily available without cost in approximately 75 percent of the cases), no longer seemed reasonable" (SSA, 1999b, p. 7). SSA further stated that this does not mean that functional ability and functional testing cannot be important components of SSA's disability evaluation process. They can and they should.

SSA has indicated that it continues to believe in the need for good information on functional ability and testing as a key part of the decision process. It has begun to focus on an alternative plan to use functional assessments in the current decision process and is addressing the issue in three different ways.

1. Ensure that functioning is appropriately considered within the current evaluation process, in terms of functional criteria that are part of SSA's standards, and in terms of providing practical policy guidelines for the use of functional testing in the current process. Many Listings include functional criteria. As the Listings are revised, SSA will seek to achieve consistency, simplicity, and administrative practicality of functional criteria.
2. Issue updated policy guidelines addressing the uses of functional assessments in the current decision process. The guidelines will address issues such as the evidentiary nature of functional assessment results, when to consider purchasing functional assessment tests, and the kind of tests to purchase.
3. SSA has asked its expert consultants to begin developing new ideas for ways to more closely investigate the use of functional testing in SSA's disability decision process (SSA, 1999b).

Committee's Assessment of SSA's Post-Redesign Plans

This section addresses the necessary prerequisites for a scientifically sound approach to disability determination at SSA. Several of the key issues that the committee had identified earlier in the study in the context of the problems associated with SSA's research plan for redesigning the disability decision process still have to be addressed with respect to the activities undertaken to improve the current process. Therefore, they bear emphasizing again in this report.

SSA has stated that the purposes of incrementally improving selected components of the current sequential disability decision process are to enhance quality of decisions, streamline the process, and update medical and vocational rules used in determining disability. This effort then, like the previous redesign effort, calls for comparative judgments. It presumes analysis of baseline information from the current decision process to assess the effectiveness of the current decision process and compare it with similar analysis of changes in the new decision process.

However, based on the information provided by SSA, the committee assumes that the agency has not conducted such baseline analysis with predetermined criteria for evaluating the components of the sequential disability decision process leading to the decision to redesign. Moreover, such analysis does not appear to have a place in the current research plan. SSA's current research approach focuses mostly on the new decision and therefore fails to build in tests that may be critical to answering the comparative questions and, ultimately, to the decision whether or not to adopt the changes in the current process.

Baseline Evaluative Criteria

Regardless of whether SSA attempts to redesign and develop a new disability decision process or leaves the current process in place and makes improvements within the individual components of the sequential process, *SSA needs to establish objective measurable criteria against which the current process can be assessed. Studies should be conducted on the existing process and data analyzed in the context of the established criteria in order to identify the nature of the problems in the current process.* Without such a capacity, proposals for "reform" may be proposals for "change," but it is impossible to determine whether they are proposals for "improvement."

When the committee reviewed SSA's research plan for the redesign initiative it was unable to conclude that SSA had put a satisfactory research plan in place (IOM, 1998). The research proposals and other documents reviewed by the committee provided no indication that SSA had conducted any baseline analysis with predetermined criteria for evaluating the Listings component, or for that matter any other component, of the sequential determination process leading to the decision to redesign the system. Moreover, such analysis did not appear to have a place in the research plan. SSA's research approach mostly focused on the new decision process and thus failed to build in tests that may be critical to answering the comparative questions and ultimately to the decision whether or not to adopt a new decision process.

The same issues appear to exist today as SSA moves toward making incremental changes in selected components of the current decision process. In the absence of evidence to the contrary, the committee assumes that the agency again has not conducted such baseline analysis leading to the current activities to improve the existing process incrementally.

For example, in its original redesign plan, SSA had proposed to replace the current Listings with an Index of Disabling Impairments that would serve to "screen-in" the obvious cases without addressing functional capacity and vocational factors. However, nowhere had SSA specified the levels of specificity and sensitivity that would be satisfactory for this Index. There was no attempt to determine whether the current Listings satisfied the goals for specificity and sensitivity overall, or the degree to which those goals were satisfied by different Listings for different body groups and different conditions. Without such baseline evaluative criteria and analysis it seemed impossible to specify either what the problems were with the current Listings or whether some redesigned Index would do a better or worse job in relation to the agency's goals for that particular screening instrument. Unfortunately, SSA appears to be going down the same path in its current efforts.

Throughout the documents reviewed by the committee relating to the redesign research, including the scope of work for the research contracts and in presentations before the committee, SSA has recognized the need to test the new disability decision process by applying standards of validity, reliability, sensitivity, specificity, credibility, and flexibility. In addition, the stated objectives of the redesign also include requirements such as simplicity in administration, consistency, accuracy, timeliness, equity of decisions at all levels, and fairness. However, to the committee's knowledge no measurement criteria have been established to test the current and the redesigned process along any of these lines.

Measurement is the process of linking abstract concepts to empirical indicators (Carmines and Zeller, 1979). Various terms are commonly used to describe measurement. For instance, to determine the extent to which a particular empirical indicator(s) represents a given concept, one can examine the reliability of the concept, that is, the reproducibility of a decision for each case within and/or across decision makers. But a process or indicator needs to be more than reliable if it is to provide accurate results. It must also be valid. Although the terms reliability and validity are often used together, they are not synonymous. A decision may be reliable but it may not be valid. Both reliability and validity reflect matters of degree. Validity represents a set of criteria by which the credibility of research may be judged. For example, it measures the degree of agreement between the disability decision and actual fact of disability. Moreover, validity has several meanings. These include construct validity; content validity; criterion validity, which includes concurrent validity and predictive validity; and study validity, which includes internal and external validity (Last,

1983). The definitions among the disciplines of logic, epidemiology, social science, and statistics do not always correspond.

Effectiveness measures the extent to which the decision process is achieving its goals and purposes (Berk and Rossi, 1990), and the concept of effectiveness must always address the issue of "compared to what?" regardless of whether it is marginal effectiveness, relative effectiveness, or cost-effectiveness. Finally, efficiency measures the results of a process in terms of resources expended and time.

The committee notes that the fields of both science and law pay unusual attention to the definition of terms. When the two fields intersect, it is particularly important to be specific about the terms used. For example, although the word "disability" is used by SSA, the actual phenomenon focuses on an attribute more narrowly defined than the inability to perform the usual activities of daily living. It refers to the inability to engage in substantial gainful activity because of a medically determinable physical or mental impairment leading to death or expected to last for at least 12 continuous months. Such distinctions are key to SSA's use of terms such as "validity," and here the legal use of the term is important to guide the scientific criteria. The appropriate construct to use in judging whether the current or the new decision process achieves its goals must take into account the legal language.

The committee, however, recognizes that the actual fact of disability may be unobservable. For example, the number of persons with disabilities under the Social Security definition will vary as a result of judicial interpretations. Each aspect of the law is subject to differing interpretations and judgments. Problems arise at each step of the sequential decision process, whether one is determining if a person has an impairment that will result in death or that is expected to last at least 12 months or, more crucially, whether by reason of that impairment a person cannot engage in SGA. The difficulty of making such a decision, however, is not the issue. Rather there is no single true answer to the question of whether a person with that impairment should be expected to engage in SGA. Simply put, a "gold standard" does not exist. Therefore, it is necessary to substitute some criterion or target and assess how well actual determinations are meeting this target. Whatever SSA chooses as a criterion or target also must be disseminated to stakeholders and decision makers as soon as possible, along with a plan for validity assessment. Only with such openness will the validity assessments be accepted when they become available.

The brief discussion of measurement terms and issues clearly demonstrates the need for SSA to specify early in the redesign effort what it means by the terms validity, reliability, sensitivity, specificity, credibility, flexibility, and all the other related terms that it uses; and how it plans to measure them, that is, what measurable criteria will be used to assess these standards vis à vis the disability decision process research and the level of sensitivity and specificity it is willing to tolerate. The same criteria should then be used to evaluate the quality of both the current process of disability determination and any prototype to be tested.

Listings of Impairments

As stated in the previous section, SSA has decided to devote its attention to updating the Listings but no timetable has been set for the completion of the various phases of this initiative. "The Listings was originally designed to highlight readily identifiable disabling impairments. Many of the Listings have since evolved into complex and highly detailed diagnostic requirements, demanding specialized medical evidence that may not be readily available from treating sources. Some, but not all, of the Listings consider the functional consequences of an impairment; however functional considerations vary significantly among the Listings" (SSA, 1994a, p. 11). The committee believes that it is indefensible that most Listings have not been reviewed and updated in more than 10 years. SSA has stated that limited staff resources, the need to address new legislative mandates during the 1990s, and the lack of adequate research on disability criteria to support Listings updates have been at least part of the problem. SSA had not made a comprehensive revision of the adult mental disorders listings since 1985 (OIG, 2000). The report of the Office of the Inspector General (OIG) states that by the late 1990s, the National Academy of Social Insurance (NASI) (Mashaw and Reno, 1996), the General Accounting Office (GAO, 1998), and the Social Security Advisory Board (SSAB, 1998) also were expressing concern that SSA was not updating the Listings regularly, but was simply extending the expiration dates for a number of years when the Listings expired (OIG, 2000). According to the OIG report, SSA staff has acknowledged that during the 1990s they did not always have the necessary research in place to support proposing revisions to the Listings or other disability projects.

SSA informed the committee in 1999 that it is correcting the situation; it has added 15 positions in the component responsible for Listings policy and has started to address some of the most critical needs. However, the committee is not aware of any attempt at this time to evaluate the currency and consistency of the Listings based on specific criteria, or at least those groups of conditions that account for a significant proportion of the disability rolls. SSA appears to have made the decision to modify certain Listings without any attempt to first evaluate them and use the findings to guide the update of the Listings.

The committee in its interim report to SSA (IOM, 1998) supported the conclusion of the Disability Policy Panel of NASI to give high priority to research related to the Listings as well as to evaluate the consistency of the presumptions underlying the Listings for different body systems (Mashaw and Reno, 1996). The committee stated that *SSA should conduct the necessary research, prior to making changes in the Listings, to (1) determine whether or not the current Listings satisfy the agency's goals for specificity and sensitivity, (2) determine whether or not these goals are satisfied consistently across the Listings for the different body groups or conditions, and (3) evaluate the options to correct the problems detected by these evaluations, as it develops any new list of medical impairments.* The committee has not changed its position.

It appears likely that the agency's agenda for reform in this area will be driven as much by internal and external anecdotal concerns, including general perceptions of which Listings are the most outdated, as by any long-range strategy. Nevertheless, the committee believes that a successful process of Listings revision must be based on a systematic approach to evaluation, design, and testing. The committee has not seen any indication of a plan for determining the specificity and sensitivity parameters for any existing or proposed Listing. Developing such parameters seems critical to both the scientific and the political validation of the Listings as a decisional tool.

Because the Listings screen is meant to be used to identify clear cases of disability, one would expect this screen to be devised such that it is highly specific (seldom identifies false positives) and relatively sensitive (identifies some substantial number of true positives). The question for SSA is how specific and how sensitive. In order to undertake meaningful research on the validity of any medical listing, SSA must be able to specify the acceptable level of specificity and sensitivity by which it can validate the screen against those criteria.

SSA provided the committee with a list of ongoing projects designed to update the Medical Listings and improve their performance. The committee, however, has no information suggesting that baseline criteria were established at the outset or that any method was developed for validating the existing and proposed new Listings against those criteria. These are serious and difficult issues. As SSA moves forward to incrementally revise and reform the current decision process, it must be able to determine whether or not changes are improving the accuracy of the process. Indeed, it has to be able to make these determinations prior to the time that changes are implemented on a national basis. Whether or not specific Listings need to be improved and the direction of that improvement must await the results of the baseline evaluation and subsequent reevaluation.

The committee in its interim report to SSA (IOM, 1998) recommended that early in the redesign effort SSA should specify how it will define, measure, and assess the criteria it will use to evaluate the current disability decision process as well as any alternatives being developed. Because of the critical importance of this issue and the assumption that not much has been done in this area, the committee reiterates its position. In any scientific process, standards of acceptance or rejection are declared before, not after, data are analyzed. Similarly in an evaluation research process, evaluative criteria and validation plans should be determined by the agency early in the research process and not, as currently planned, after the Listings are identified for updating and changes are developed.

Recommendation 6-1: The committee recommends that prior to making the changes in the current decision process, SSA should

a. **establish evaluative criteria for measuring the performance of the decision process;**
b. **conduct research studies and analyses to determine how the current processes work relative to these preestablished criteria, and**
c. **evaluate the extent to which change would lead to improvement.**

Analysis of data from such studies in the context of the established criteria would identify the nature of the gaps between what the program is supposed to achieve and its actual performance. Without these research steps and analysis, there is no objective way to conclude if the changes are more effective and more efficient than the existing process.

Since SSA is devoting its attention to updating the Listings, this recommendation is most applicable at this time to the Listings. However, the committee notes that the Listings apply only to one step (step 3) of the five-step sequential evaluation process for determining disability. The baseline evaluation recommended for the Listings should ultimately evaluate the total process and not just one component.

The committee is encouraged to note the recent cooperative agreement awarded by SSA in December 2000 to the Disability Research Institute (DRI) to undertake research for developing a process of validation of the Listings in order to assess them and to ensure that changes made actually result in improvements in the disability decisions. This research will (1) compile and list published articles in the literature pertaining to the validation of disability insurance decisions, (2) undertake a critical review of the literature on assessing validity and, primarily and most importantly, on the development of appropriate validation criteria, and (3) review methods by which these validation criteria could be operationalized. Although completion and implementation of this project may help validation efforts for future revisions of the Listings and other components of the disability determination process, unfortunately no such input exists for the revisions currently under way or completed.

The committee recognizes that SSA faces two major challenges in providing an appropriate research base for the improvement of the disability decision process. First, it operates an ongoing program that requires continuous incremental adjustment in order to make appropriate decisions. Second, the environment of disability decision making is constantly shifting in ways that have unanticipated consequences for the current process and that generate movements for substantial reorientation of the entire disability benefit programs. Moreover, it goes without saying that SSA should recognize the cost trade-offs when it sets targets for sensitivity and specificity, and other measures.

Assessing Vocational Capacity

Determining the Demands of Jobs in the National Economy. As indicated in the previous section, the Dictionary of Occupational Terminology (DOT) is no longer being updated by the Department of Labor, leaving SSA with no replacement. The DOT has served as a primary tool for determining whether a claimant has the capacity to work. The Department of Labor is replacing DOT with O*NET. The committee, in its preliminary assessment of the redesign research plan, had expressed its concerns that O*NET as it was being developed for DOL would not meet SSA's needs (IOM, 1998). For one thing, it focuses on average rather than minimum requirements as needed by SSA. The committee also questioned how SSA planned to supplement O*NET with respect to contextual and other factors that are not well covered. Discussions at the workshop sponsored by the committee on *Measuring Functional Capacity and Work Requirements* (IOM, 1999a) pointed out the problems associated with using O*NET for SSA's purposes. The DOL expects to use O*NET, as a comprehensive database of work requirements for use in job training, job counseling, and job placement for the department's employment and training programs and for use by individual state Employment Security Agencies in the extensive work that they do with workers who need jobs or who have recently become unemployed.

As discussed at the workshop, although O*NET is very useful for DOL's purposes, SSA's purpose in defining the functional capacity to work for purposes of the disability legislation is very different from the purposes of the DOL in creating O*NET. SSA's purpose is much more difficult. Moreover, the labor market and occupational literature indicate that there are many difficult measurement problems related to occupation and job characteristics. Information developed by job incumbents is not always consistent with the information developed by job analysts, and the information developed by job analysts is not always consistent with the views of workers' supervisors. The Bureau of Labor Statistics conducts employer surveys that try to define the characteristics of a job that affect its pay levels, but even there measurement difficulties sometimes exist. In addition, from the perspective of the worker—as with a disabled individual—it is often a bundle of capabilities that the worker brings to the job that makes the work experience a success or a failure.

Workers with the same educational backgrounds have different skills, work ethics, and orientations to work. These in turn bring a different bundle of capabilities to a job, and their performance is affected by those capabilities. In addition, the task of developing a set of factors that capture the essence of each occupation that makes practical sense is complex and difficult. Clearly, a great deal more careful research and experimentation is required to evaluate what functional capacity to work really means and exactly how it would be applied to persons with disabilities.

When the committee reviewed SSA's redesign research plan, there were no indications in the plan that the gaps in O*NET will be carefully considered and no specific research to fill those gaps was identified. The committee, therefore, had recommended that SSA should develop an interim plan for an occupational information classification system until a more permanent solution is found. The committee also suggested that SSA enter into an interagency arrangement with the DOL to initiate a version of O*NET that would collect information on minimum, in addition to average, job requirements to better serve SSA's needs to assess ability to engage in SGA.

Subsequent to the committee's interim report, SSA revised the work requirements of its original task order contract on integration, synthesis, and development of the redesign process to focus solely on a comprehensive assessment of O*NET as a replacement data source for the current decision process. SSA did not necessarily expect this work to produce a comprehensive resolution to the problem. It believed, however, that it must complete such an analysis to move forward (SSA, 1999b).

The final report of the contractor (AIR, 2000) identified several positive and negative aspects associated with O*NET's incorporation into SSA's disability determination process. Some of the positive aspects identified include the ability to (1) obtain consensus from a variety of sources on the set of 54 O*NET descriptors appropriate for use in SSA's disability decision process; (2) ascertain the number and identity of occupational units that are represented at various intervals of the descriptors' rating scales; (3) determine if the occupational units are sufficiently representative for SSA's use; (4) identify 33 occupational units that contain at least one sedentary or unskilled job, as defined by DOT, representing approximately 3 percent of the 1,122 occupational units in O*NET; and (5) provide excellent descriptions of occupations in the task lists that decision makers can use to help determine the specific activities that comprise the claimant's current or previous occupations.

On the other hand, several aspects of the O*NET structure and content could lead to problems if SSA incorporated O*NET into the decision process. More than half of the occupational units had at least one domain for which the majority of descriptors were unreliable. The final report (AIR, 2000) emphasizes that a major overarching problem with O*NET is the numerical ratings. These ratings do not seem to be consistent across occupational units. The contractor's analysis found that the ratings of more than half of the descriptors are unreliable. Moreover, the DOT titles are grouped by dimensions that are unrelated to worker characteristics or requirements of the O*NET descriptors. Several of the 54 selected descriptors contain O*NET ratings with interrater reliabilities lower than .70. The report concludes that the numerical ratings on O*NET descriptors, and therefore on any O*NET occupational unit, underlie the problems of O*NET. Therefore, SSA must exercise extreme caution in drawing inferences about the relation between specific numerical values on a rating scale and specific level of required functioning. The report further states that the foregoing concerns provide sufficient evidence to warrant SSA's careful consideration of the quality of either analyst or incumbent ratings as conducted and proposed for O*NET. The report also suggests that O*NET's descriptor data may not be as precise as they seem, resulting in measurement errors as well as improper interpretation of the severity of claimants' impairments (AIR, 2000).

Without an appropriate characterization of job requirements that can be matched to the vocational characteristics of disability claimants, SSA might be cast back into the era in which it relied extensively on the testimony of "vocational experts," or their written evaluations, as the way to integrate claimants' functional capacities, vocational factors, and the demands of work into an objective determination of their capacity to engage in substantial gainful employment. Barring some resolution, SSA will be left with no objective basis upon which to justify decisions concerning an individual's capacity to do jobs in the national economy (AIR, 2000).

SSA realizes that O*NET will not work for its needs without major reconstruction of the system. SSA is taking steps toward resolving the problems. The committee is informed that SSA has reopened its dialogue with DOL to explore other ways of incorporating information about the requirements of work into the decision process and is actively pursuing with DOL the issue of an occupational database on a national level to avoid two separate databases with separate funding. It is also meeting with the various associations of rehabilitation specialists, occupational and physical therapists, and workers' compensation analysts. Private sector stakeholders have organized an interdisciplinary task force. It plans to meet with SSA and DOL to decide what is needed and how best to go about getting the information.[2]

Age, Education, and Work Experience. SSA has not updated the research base on the effect of age, education, and work experience on work disability that had been used in developing the medical–vocational guidelines of 1978, known as the "grid rules." Since then, much has changed with regard to the relative importance of each of these factors. As part of the initiative to redesign the decision process, SSA included in its research plan the evaluation of the effect of vocational factors—age, education, and work experience—on the ability to work or adapt to work in the presence of functional impairment. To assist in deciding an appropriate way to incorporate into the redesigned disability decision process the specific statutory requirement to consider an individual's vocational factors in determining ability to work, SSA entered into a reimbursable agreement with the Federal Research Division of the Library of Congress to review and evaluate published literature and any other research pertaining to the subject. A report of the review was submitted to SSA in 1998. The findings of the review are summarized very briefly below.

Although *age* strongly affects decision making under the vocational grid, the literature review of the existing literature suggests that age may have little or no independent influence on ability to work (as distinguished from the likelihood of being hired or retained). Rather than chronological age being a common contributing factor to declining capacities, current studies suggest that the population actually becomes much more heterogeneous with respect to its functional capacity as it ages. Moreover, except for the relatively vague concept of "adaptability," age does not seem to have a strong correlation with modal ability to work.

Education is clearly an important factor in employability. It affects the ability to acquire new skills, and earning power is related to education level. It is especially a problem with mental impairments. However, education above basic literacy levels has an uncertain relationship to the ability to do jobs that would produce substantial gainful employment. High levels of education are not necessary for jobs paying $8,400 per year. High levels of education may, of course, suggest that only the most debilitating injuries or illnesses would prevent substantial gainful employment by persons with such levels of educational attainment. In combination, these attributes suggest that education may be important as a vocational factor only at the upper and lower range of educational attainment, but not in the middle ranges.

---

[2]Personal communication, Sylvia Karman and David Barnes, Office of Disability, SSA, October 3, 2001, and December 4, 2001.

Increasingly basic skills are a priority with employers in identifying new employees. Employers value basic skills that flow from education, including the capacity to learn new skills or information, much more than they value job-specific skills. As with age, the independent influence of work experience is difficult to evaluate. Work experience is certainly important in terms of capacity to return to a worker's own occupation after an injury or illness. However, the vocational factors are used most often in evaluating the capacity of workers to do jobs other than those that they have held before.

In summary, the review raised questions about the utility of multiple gradations of educational attainment in evaluating the vocational factors in disability determination and the utility of making determinations based on a worker's transferable skills. Existing knowledge concerning vocational factors and their impact on the ability to perform jobs in the national economy raises challenging questions about the continuing validity of the approach taken by SSA's existing grid rules. It suggested a critical need for a program of research designed to validate or reform the use of vocational factors in SSA's disability decision process.

SSA recognizes that it may have to make significant revisions to the rules it uses to determine disability, especially in light of the changes that the Department of Labor is making in its occupational data. SSA's current rules, especially the grid rules, are based in part on both the organizational structure and the data content of the DOT. Without it, those rules will probably have to be revised in a fundamental way. SSA also recognizes that such a revision might also necessitate a change in the way it incorporates evaluation of age, education, and work experience in its disability decision-making process.[3]

Variations in Disability Allowances. As shown in the previous chapters, over the past two decades the number of disability beneficiaries as a share of the civilian labor force has risen steadily. Although applications for benefits have increased only moderately, the number of new beneficiaries has nearly doubled. Disability allowance rates (awards as a percentage of applications) have varied over time from 31.4 percent in 1980 to nearly 47 percent in 2000 (see Table 2-1). Variations in allowance rates occur for several reasons. For example, SSA's standards for judging claims differ over time. Dramatic reductions in approval rates occurred when standards were abruptly tightened in 1980 and then subsequently made liberal. Significant differences are observed in approval rates across states, between the state Disability Determination Service (DDS) decision mak-

---

[3]Personal communication, David Barnes, Office of Disability, SSA, December 14, 2001.

ers, and between DDSs and administrative law judges (ALJs). The approval rate is also influenced by legislative changes as well as court decisions. The adequacy of resources to process and review cases also affects the disability allowance rates.

Increased research is needed to explain the variation in the rates at which applications for disability benefits are approved, why these changes take place, and whether they are predictable. To what degree are the growth and changes in disability allowances related to societal factors, and to what degree have they been influenced by changes in the program rules and operations? Such research should involve examination of the disability decision-making processes and the standards applied by SSA, the differences among states, and the differences among DDSs and the ALJs and among states. For instance, in the short run, changes in the prevalence of impairments are not as likely as changes in the way SSA evaluates the various impairments. Untangling the effects of the demand side (the growth in the number of SSDI and SSI applicants) and the supply side (the SSA disability decision processes) and prescribing remedies is difficult, but careful research in these areas will help shed light on this comparatively neglected area.

Recommendation 6-2: The committee recommends that the Social Security Administration conduct research on

a. **improving the ability to identify and measure job requirements for the purpose of determining work disability;**
b. **investigating the role and effects of vocational factors in the disability decision process; and**
c. **understanding reasons for variations in allowance rates among states and over time.**

ALTERNATIVE APPROACHES TO THE CURRENT DECISION PROCESS

The objectives of the current disability decision process appear to be an attempt to make accurate decisions about the capacity to engage in substantial gainful employment consistent with the statutory definition of disability as consistently, expeditiously, and cost-effectively as possible within a system that is hierarchically accountable and makes determinations at a relatively low cost. Mashaw (1983) refers to it as "bureaucratic rationality." Although interpretive judgment is clearly necessary in adjudication, the process has become an increasingly rule-bound system that strives, with greater or less success, to decide similar cases in a consistent manner, in accordance with the statutory definition of disability, and in a timely and efficient manner.

Bureaucratic rationality, however, is not the only model of an adjudicatory process that might be applied to disability benefits determination. Some parts of the process, in particular decision making at the ALJ level, are more like an adversary adjudicatory process. Moreover, one could imagine the process as one that looks like the Internal Revenue Service system in which "claims auditors" might have the capacity to grant, deny, or even "settle" claims and would then defend those decisions at subsequent levels of review. Workers' compensation and unemployment compensation systems provide other examples of adversary models of benefits adjudication.

Another, radically different approach would conceive of disability benefits designed to assist claimants in receiving appropriate medical attention and vocational rehabilitation as well as appropriate income supports. In this model the basic goal of the program would be to move claimants back toward productive work and to use benefits both as a means to facilitate the return to work process and as an ultimate fallback for those claimants whose impairments make continued work impossible. This is the approach used by many private disability insurers who manage employment-based disability plans in the United States, and it is the dominant model in certain foreign systems, such as those in Sweden and Germany.

Concern also is expressed that environmental factors, including environmental barriers to work are not taken into consideration in defining work disability.

Recent legislation makes clear that Congress is increasingly interested in the "return-to-work" model and is prepared to have SSA experiment with some alternative strategies that might facilitate the pursuit of work rather than benefits. The Ticket to Work and Work Incentives Improvement Act of 1999 (P.L. 106-170) was signed into law on December 17, 1999. One major provision of the law establishes the Ticket to Work and Self Sufficiency Program, or Ticket Program. This provision provides that eligible SSDI and SSI beneficiaries with disabilities will receive a ticket (or voucher) they can use to obtain employment services, vocational rehabilitation services, or other support services from an approved provider of their choice. The law also expands Medicaid and Medicare coverage to more people with disabilities who work.

Recommendation 6-3: The committee recommends that the Social Security Administration initiate a research program for testing decision process models that emphasizes rehabilitation and return to work.

In conclusion, although SSA has deferred a major redesign of the disability decision process, the committee believes that it is paramount that the determination of disability not only be timely, understandable, straightforward, and feasible, but also provide accurate and consistent decisions that are fair to the claimant and to the government. To this end the committee believes that SSA should undertake a systematic, long-term program of research—intramural and extramural—that provides baseline information on all key aspects of the current disability decision process and subsequent evaluative data on all future change aimed at improving the effectiveness and efficiency of the work disability determination process currently in use in the United States.

7

Enhancing SSA's Research Capacity

During the nearly six years that this committee met, it reviewed a large number of research reports, journal articles and government reports, and relevant internal documents and other unpublished reports provided by the Social Security Administration (SSA). It also heard presentations from SSA staff on various aspects of their work and progress made on the projects reviewed. Experts in the field addressed the committee during its meetings and also participated in two large workshops organized by the committee.

The committee's recommendations in the preceding chapters and in its interim reports to SSA are intended to better inform public policy by developing a national data system for monitoring on a continuous basis the size and characteristics of the population eligible for Social Security disability benefits and enhancing research leading to improved assessment of work disability for the purpose of awarding benefits. The committee has recommended major research efforts, including research on the measurement of work disability in a survey context, evaluation of the role of the environment and vocational factors in determining work disability, evaluation of the functional capacity of applicants for disability benefits, and testing decision process models that emphasize rehabilitation and return to work. Such research cannot be accomplished without appropriate infrastructure and resources, in terms of both dollars and recruitment of qualified researchers. In the course of its study the committee noted several problems related to infrastructure and research capacity in SSA and going beyond a specific individual unit of SSA or the specific subject matter under consideration. The successful reform of a disability decision process and the implementation of the national disability monitoring system depend on the resolution of these problems. The committee recognizes that the recommended enhancements would require substantial additional funds and qualified staff. This concluding chapter briefly addresses those issues.

RESOURCE REQUIREMENTS

As shown in the previous chapters, the number of disabled workers receiving Social Security Disability Insurance (SSDI) benefits and Supplemental Security Income (SSI) based on disability, as well as the costs of these programs, has grown substantially since the beginning of the programs. Continued growth is projected as the baby boom generation reaches the age of increased likelihood of disability. At the same time, disability policy has become more complex. Extensive research is needed to understand, estimate, and forecast growth to inform and guide public policy.

Over the years staff of the Office of Research, Evaluation and Statistics (ORES) in SSA has conducted a variety of excellent surveys and studies. Establishing and maintaining high-quality and relevant data systems for appropriate analysis and dissemination requires a sufficient and capable intramural research staff. The committee finds that there has been a loss of survey design and analytical capacity at the very time such work needs to be expanded. In the past two decades, downsizing has adversely affected both the ORES and the disability program (Institute for Health and Aging, 1997). A lesson learned from the experience with the National Study of Health and Activity (NSHA) is the importance of staffing to handle the issues that are critical in launching a large complex survey. The current impoverished research capability in SSA not only affects the timely analysis of data collected, but also leads to inability to anticipate important issues and respond to them. If not corrected, this situation will impair the ability of SSA to meet its policy needs in the twenty-first century.

The committee notes the limited resources allocated to all Social Security research activities. Two recent reports of the Social Security Advisory Board (SSAB, 1997, 1998) also noted the very small number of staff positions and budget amounts devoted to research and recommended that SSA increase its intramural and extramural research activities. A third report (Institute for Health and Aging, 1997) reviewed the mission, resources, and capabilities of SSA's Office of Research, Evaluation, and Statistics and recommended that at least 50 new full-time positions be added to the ORES staff to strengthen the internal research and evaluation capacity, to develop and support external resources for research, and to ensure adequate funding to support these programs. While these recommendations encompass all of SSA's research activities and go beyond research in the disability program, the committee recognizes the need to revitalize and strengthen the research program of ORES and to encourage collaboration with other federal agencies in activities relevant to SSA. The committee fully endorses the recommendations made in these reports for increased resources and in-house capacity for research and commends SSA for its recent efforts to increase staff resources and research activities.

In response to these recommendations, SSA took some steps to increase staff levels in ORES from 132 positions in 1997 to an estimated 141 positions by 2000. Of this number, 99 in 1997 and 111 in 2000 were allocated to research evaluation and statistics. The remainder are distributed among publication activities, technological infrastructure, and management, administrative, and clerical functions. Clearly much more is needed to meet the demands for research and statistics in the coming years. SSA should ensure that an optimum mix of disciplines is represented on its staff. Some examples include survey methods, sampling statistics, economics, operational research, demography, epidemiologists, sociologists, cognitive psychologists, medicine, and the like.

Recommendation 7-1: The committee recommends that the intramural staff for disability research and statistics should be substantially expanded and diversified to implement the recommendations in this report.

In addition to the need for an expanded intramural research program, the committee believes that there is a major role for extramural research. A balanced program of intramural and extramural research is needed. "No amount of external research will replace the need for the agency to invest in the internal research capability, for it is essential in itself and inextricably linked with the capacity to implement and use an effective extramural program" (Institute for Health and Aging, 1997, p. 29). Moreover, an extramural research program places its own demands on the agency's research staff. Oversight responsibility rests with the agency for careful evaluation of the work of external researchers to ensure the quality, adequacy, and appropriateness of the products, and for designing the approaches to testing and experimentation.

In addition, SSA has research grant authority under Section 1110 of the Social Security Act. Over the years, this authority has been the basis for a relatively small research grant program that has been managed by ORES. Grants were solicited for research in targeted areas, and in addition, investigator-initiated grants were peer reviewed and awarded. The funding of this research grant program has been erratic, with no funds allocated to the program during the past three decades.

Peer-reviewed extramural research programs have proved highly successful in the field of health services and clinical research. The Centers for Medicare and Medicaid Service (formerly the Health Care and Financing Administration), the Agency for Health Care Research and Quality, the National Institutes of Health, and the National Science Foundation have developed highly successful and sophisticated systems for review of investigator-initiated research in a wide variety of health areas. A similar, strong program is needed in the social insurance area and should be operated and managed by high-level qualified professional staff in ORES.

Recommendation 7-2: The committee recommends that the Social Security Administration (SSA) expand and diversify its extramural research program to include a balance of contracts, cooperative agreements, and investigator-initiated grants. This broadened research program would prepare SSA for the anticipated growth in the demands on the disability programs and help bring about the needed fundamental changes in its disability programs.

The committee notes, however, that although the grant authority has been unfunded in recent years, SSA has taken some steps in that direction by awarding cooperative agreements. Lacking adequate infrastructure at this time to operate an effective grant program, cooperative agreements with less demanding infrastructure could begin to serve some of the purposes similar to investigator-initiated research. Two such large agreements are the Disability Research Institute described earlier and the Retirement Research Consortium (RRC). These consortia draw researchers from several universities together. Their main goals are to foster research and evaluations, dissemination of information on retirement, and other SSA-related social policy including disability policy, training and education, and facilitating the use of SSA's administrative data by outside researchers. To meet these goals, the centers perform many activities including research projects, policy briefs and working papers, annual conferences, and training. The RRC currently is composed of two, university-based, multidisciplinary centers, administratively based at Boston College and the University of Michigan.

SSA should view the ability to fund intramural research, external research—contracts, cooperative agreements, and grants—as separate tools to improve the functioning of the agency. Each can offer SSA leadership unique ways to learn of causes of external social and economic phenomena that affect the applicant pool to SSDI, influences on how individuals make decisions about application to SSDI, the effectiveness of the processing of applications, the dynamic nature of eligibility, and what influences return to work among those eligible for SSDI. By judicious coordination of the three programs of research, SSA can greatly enhance management intelligence needed for assessing the desirability of change in policies.

Intramural research can most effectively be focused on internal information analysis, studying the effectiveness of administrative procedures in the SSDI program. In addition, intramural researchers can be statistical analysts of external data used to estimate key demand statistics for SSDI services. Finally, intramural researchers should supply key analysis of direct utility to SSA's policymakers.

Research contracts are effectively used to collect well-specified data using standard techniques. For example, contracts might be used to provide ongoing estimates of key statistics of interest to SSA, collect data on an ongoing basis, or providing ongoing statistical support services. The value of a research contract is the assurance of quality and cost efficiency for ongoing work.

Cooperative agreements are best used when SSA has identified well-defined research products but there may be uncertainties about how best to obtain those products. With a cooperative agreement, as implied by the name, SSA staff can interact with external researchers to help shape methods and products throughout the agreement by working as a partner with these researchers. Thus, cooperative agreements seek new ideas from outside the agency for research information that has, at least, been sketched out prior to the agreement.

Research grants offer the greatest opportunity for innovative ideas but provide for little control by SSA management. They are reviewed by a set of peer scientists outside the agency. They are evaluated by the soundness of scientific thinking motivating them and the likelihood of advancement of understanding of problems facing SSA. SSA would define sets of key questions that it wanted to be addressed through the grant mechanism. Proposals would be initiated by external researchers. In comparing grants and cooperative agreements, grants are probably best used for high-risk, but high-payoff, domains of knowledge. If SSA exercised the grant mechanism, it is likely that real breakthroughs in the understanding of key population phenomena may be possible over time. These are the types of findings that could lead to new ways of administering the programs or new programs.

NEED FOR FUNDAMENTAL CHANGES

The previous chapters of this report make abundantly clear that SSA has been given a difficult task and dwindling resources to deal with it. The situation will get worse and not better in light of the anticipated growth in demands on the program as the baby boom generation reaches the age of increased likelihood of disabilities. In its recent reports the SSAB (2001, 2002) has reached similar conclusions and has recommended major rethinking of the disability program.

Little doubt exists that the current system is in need of improvement. It needs better understanding of the prevalence of disability in the population and the characteristics of this population, and better information about the job market, and about qualifications for jobs. The Department of Labor's Dictionary of Occupational Titles (DOT) is no longer being updated, and as of now SSA has no replacement for the DOT, leaving a critical vacuum. The problems of Social Security's disability decision process are deep and fundamental. This is not adequately reflected in the agency's research agenda. Making small changes with the current system may not resolve the basic problems. Changes to O*NET (the Occupational Information Network), even if they are feasible, updated Listings of Impairments, and the like may help but will not necessarily solve the basic problems facing SSA. While the Listings can and should be updated in light of the changes in medical knowledge, methods to validate them are not yet in place. They need updating, however, even if we have no perfect instrument for their validation. Moreover, attempts to validate them will be confronted with the stark fact that so many persons who meet the Listings work at normal jobs in the national economy.

SSA must recognize that the present system for determining program eligibility may not be sustainable in the future and that it must think about different orientations and different ways in which the task of making these decisions is accomplished. It needs to have some mechanism to systematically give thought to these issues and initiate appropriate research. For example, SSA should initiate research on the costs and benefits of the current decision process and alternate innovative approaches. Without sufficient resources, however, SSA cannot accomplish this forward-looking agenda.

SSA recognized these problems in the early 1990s when it decided to rethink and fundamentally redesign the disability decision process. It stated that "the fragmented nature of the disability process is driven by and exacerbated by the fragmentation in SSA's policymaking and policy issuance mechanisms. Policymaking authority rests in several organizations with few effective tools for ensuring consistent guidance to all disability decisionmakers. Different vehicles exist for conveying policy and procedural guidance to decisionmakers at different levels in the process. . . . The organizational fragmentation of the disability process creates the perception that no one is in charge of it. . . ." (SSA, 1994a, p. 11)

The SSAB (2001) also concluded that the disability policy and administrative infrastructure are weak and that constructive change and additional resources are required. It stated that "the problems with the administrative infrastructure begin at the top, where SSA's current organizational structure diffuses responsibility over nearly every component of the agency. They continue throughout the disability system, where a fragmented and uncoordinated administrative arrangement makes consistency and fairness in decisionmaking difficult to achieve."

"Problems in the area of policy are equally critical. For many years, disability policy has tended to be guided by court decisions and other pressures rather than by a well-thought-out concept of how the programs should be operating. Policy is articulated by too many voices with no single source of policy to which decisionmakers can turn for guidance and direction. Moreover it is inconsistent with the objectives of many disabled individuals to participate in the economic mainstream through employment" (SSAB, 2001, p. 29). In that report SSAB concluded that ". . . the issues facing the disability programs cannot be resolved without making fundamental changes. In our view these changes must be evaluated within the context of clear goals and objectives. . ." (p. 11).

The committee endorses the conclusions reached by SSA and the SSAB that underscored the need for fundamental change in the Social Security disability programs. SSA desperately needs a long-term, systematic research program to inform and guide (a) the anticipated growth in demands on SSA's disability programs, and (b) improvements in the disability determination process.

In conclusion, the committee commends the SSA for initiating the daunting tasks of developing a national survey to improve the information base needed for monitoring and projecting the size and characteristics of the eligible population for guiding disability policy, and of attempting to overhaul the disability decision process to focus directly on developing new ways to assess the applicant's functional ability or inability to work as a consequence of the medical impairment and to rely on these standardized functional measures to reach decisions. The ultimate goals of such a redesigned system were to make it simple to administer, to facilitate consistent application of rules at each decision level, to provide accurate and timely decisions, and to be perceived by the public as straightforward, understandable, and fair.

Although during the course of its study the committee identified much that needed changing, and it continues to be concerned about some of the decisions made by SSA, it recognizes, and is pleased, that SSA made many modifications in response to its recommendations for improving the National Survey of Health and Activity. *The committee believes that the blueprint for action it recommends for developing and implementing a disability monitoring system for Social Security programs, and for needed research relating to the redesign of the disability decision process will contribute toward a significantly improved and efficient system of measuring and monitoring work disability that will better inform public policy and serve the public. This blueprint is worthy of full funding and adequate staffing support by both the Congress and the executive branch of government.*

References

Acemoglu D, Angrist J. 1998. *Consequences of Employment Protection? The Case of the Americans with Disabilities Act.* NBER Working Paper No. w6670. National Bureau of Economic Research. Cambridge, MA.

ADA (Americans with Disabilities Act). [Online]. Available: http://www.usdoj.gov/crt/ada/pubs/ada.txt. [Accessed August, 2001.]

Adler M. 1996. Federal disability programs. *Encyclopedia of Financial Gerontology.* Institute for Socio-Financial Studies. Middleburg, VA.

AIR (American Institutes for Research). 2000. *Synthesis, Integration, and Completion of Research into a New Disability Decision Making Process and Development of an Initial Prototype of That Process.* Prepared under contract with SSA (Contract No. 0440-97-32258). Washington, DC.

AMA (American Medical Association). 1993. *Guides to the Evaluation of Permanent Impairment.* 4th ed. Chicago: American Medical Association.

Autor D, Duggan MG. 2001. *The Rise in Recipiency and the Decline in Unemployment.* NBER Working Paper No. w8336. National Bureau of Economic Research. Cambridge, MA.

Badley EM. 1993. An introduction to the concepts and classifications of the International Classification of Impairments, Disabilities, and Handicaps. *Disability and Rehabilitation* 15:161–178.

Badley EM. 1995. The genesis of handicap: Definitions, models of disablement, and role of external factors. *Disability and Rehabilitation* 17:53–62.

Berk RA, Rossi PH. 1990. *Thinking About Program Evaluation.* Newbury Park, CA: Sage Publications.

Berkowitz E. 1997. The historical development of social security in the United States. In: Kingson ER, Schulz JH, eds. *Social Security in the 21st Century.* New York: Oxford University Press.

Berkowitz M, Burton J. 1987. *Permanent Disability Benefits in Workers Compensation.* W.E. Upjohn Institute.

Birren J, Dieckermann L. 1991. Concepts and content of quality of life in the later years: An overview. In: Birren J et al., eds. *Quality of Life in the Frail Elderly*. New York, NY: Academic Press 344–360.

BLS (Bureau of Labor Statistics). January 1999. *Employment and Earnings, 1998*. U.S. Department of Labor, Washington, DC.

Bound J. 1989. The health and earnings of rejected disability insurance applicants. *The American Economic Review* 79(3):482-503.

Bound J, Waidmann T. October 2000. *Accounting for the Recent Declines in Employment Rates Among the Working-Aged Disabled*. Cambridge, Massachusetts: National Bureau of Economic Research Working Paper 7975.

Bound J, Waidmann T. June 2001. Accounting for recent declines in employment rates among working-aged men and women with disabilities. Paper presented at the Cornell Employment and Disability Policy Institute conference, *The Persistence of Low Employment Rates for People with Disabilities: Causes and Policy Implications*. October 2001. Washington, DC.

Burkhauser R, Daly M, Houtenville A. January 2000. How working age people with disabilities fared over the 1990s business cycle. *Rehabilitation Research and Training Center for Economic Research on Employment Policy for Persons with Disabilities*. Cornell University. Ithaca, NY.

Burkhauser R, Daly M, Houtenville A, Nargis N. March 2001. Economic outcomes of working-age people with disabilities over the business cycle: An examination of the 1980s and 1990s. Paper presented at the Cornell Employment and Disability Policy Institute conference, *The Persistence of Low Employment Rates for People with Disabilities: Causes and Policy Implications*. October 2001. Washington, DC.

Carmines EG, Zeller RA. 1979. *Reliability and Validity Assessment*. Newbury Park, CA: Sage Publications.

CDC (Centers for Disease Control and Prevention). 1988. *CDC Surveillance Update*. Atlanta, Georgia.

Census (Bureau of the Census) and BLS. 2000. *Current Population Survey: Design and Methodology. Technical Paper No. 63*. U.S. Department of Commerce and U.S. Department of Labor, Washington, DC.

Cohen M. 1957. *A Preface to Logic*. New York: Meridian Books.

Cohen S. 1999. (Submitted for publication to Medical Care) *Design and Estimation Strategies and Innovations in the Medical Expenditure Panel Survey for the Measurement of Health Care Expenditures and Quality*. Agency for Health Care Research and Quality, Rockville, MD.

DeLeire T. 2000a. The unintended consequences of the Americans with Disabilities Act. *Regulation* 23(1):21–25.

DeLeire T. 2000b. The wage and employment effects of the Americans with Disabilities Act. *Journal of Human Resources* XXXV(4):693–715.

DHHS (U.S. Department of Health and Human Services). 1982. *Implementation of Section 304(g) of Public Law 96-265, Social Security Disability Amendments of 1980*. Report to the Congress by the Secretary of Health and Human Services. Washington, DC: DHHS, Social Security Administration.

DHHS. 1992 (unpublished). *The Social Security Disability Insurance Program: An Analysis*. Report of the Department of Health and Human Services. Washington, DC.

Fleishman EA. 1972. Structure and measurement of psychomotor abilities. In: Singer RN, ed. *The Psychomotor Domain: Movement Behavior*. Philadelphia: Lea and Febiger.

Fleishman EA. 1999. Linking components of functional capacity domains with work requirements. In: Wunderlich GS, ed. *Measuring Functional Capacity and Work Requirements. Summary of a Workshop*. Washington, DC: National Academy Press

Fougeyrolles P. 1995. Documenting environmental factors for preventing the handicap creation process: Quebec contribution to ICIDH and social participation of people with functional differences. *Disability and Rehabilitation* 17(3/4):145–153.

Fougeyrollas P, ed. 1998. *ICIDH and Environmental Factors International Network.* Vol. 9, Nos. 2–3. Quebec, Canada: Canadian Society for the ICIDH.

Frey WD. 1984. Functional assessment in the '80s: A conceptual enigma, A technical challenge. Chapter 1 in *Functional Assessment in Rehabilitation.* Halpern AS, Fuhrer MJ, eds. Baltimore, MD: Paul H. Brookes Publishing Company.

Friedman S, Wachs T. 1999. *Measuring Environment Across the Life Span: Emerging Methods and Concepts.* Washington, DC. American Psychological Association.

Gallicchio S, Bye B. 1980. *Consistency of Initial Disability Decisions Among and Within States.* Washington, DC: U.S. Government Printing Office, (DHHS/SSA): SSA Pub. No. 13–11869

GAO (U.S. General Accounting Office). 1994. *Social Security Disability: SSA Quality Assurance Improvements Can Produce More Accurate Payments.* Report to the Chairman, Committee on Finance, U.S. Senate. Pub. No. GAO/HEHS–94–107. Washington, DC: GAO.

GAO. 1995. *Social Security Disability: Management Action and Program Redesign Needed to Address Long-Standing Problems.* Testimony by Jane L. Ross to the Subcommittee on Social Security, Committee on Ways and Means, House of Representatives. Pub. No. GAO/HEHS–95–233. Washington, DC: GAO.

GAO. 1996. *SSA Disability Engineering: Project Magnitude and Complexity Impede Implementation.* Testimony by Diana S. Eisenstat to the Subcommittee on Social Security, Committee on Ways and Means, House of Representatives. Pub. No. GAO/HEHS–96–211. Washington, DC: GAO.

GAO. 1997a. *Supplemental Security Income: Long-Standing Problems Put Program at Risk for Fraud, Waste, and Abuse.* Testimony by Jane L. Ross to the Subcommittee on Oversight, Committee on Ways and Means, House of Representatives. Pub. No. GAO/HEHS–97–88. Washington, DC: GAO.

GAO. 1997b. *Social Security Disability: SSA Must Hold Itself Accountable for Continued Improvement in Decision-Making.* Report to the Chairman, Subcommittee on Social Security, Committee on Ways and Means, House of Representatives. Pub. No. GAO/HEHS–97–102. Washington, DC: GAO.

GAO. 1998 *SSA's Management Challenges: Strong Leadership Needed to Turn Plans into Timely, Meaningful Actions.* Testimony by Jane L. Ross to the Subcommittee on Social Security, Committee on Ways and Means, House of Representatives. Pub. No. GAO/HEHS-98-113. Washington, DC: GAO.

Groves RM. 1989. *Survey Errors and Survey Costs.* New York: John Wiley.

Gustafson S, Rose A, Mulqueen C, Matheson N, Michel R, Bott C. 2000. *Synthesis, Integration, and Completion of Research into a New Disability Decision Making Process and Development of an Initial Prototype on That Process.* American Institutes for Research, Washington, DC.

Haber L. 1971. Disabling effects of chronic disease and impairment. *Journal of Chronic Diseases* 24(7):469–487.

Hale TW. 2001. The lack of a disability measure in today's Current Population Survey. *Monthly Labor Review,* 38–40.

Helmick C, Lawrence R, Pollard R, Lloyd E, Heyse S. 1995. Arthritis and other rheumatic conditions: Who is affected now, who will be affected later? *Arthritis Care and Research* 8:203–211.

Hu J, Lahiri K, Vaughan D, Wixon B. 1997. *A Structural Model of Social Security's Disability Determination Process.* ORES Working Paper Series, No. 72 Ed. Washington, DC: SSA, Office of Research, Evaluation, and Statistics.

Institute for Health and Aging. 1997. *Strengthening Policy Development Work Within the Social Security Administration: A Review of the Mission, Resources, and Capabilities in the Office of Research, Evaluation, and Statistics.* San Francisco: University of California.

IOM (Institute of Medicine). 1991. *Disability in America: Toward a National Agenda for Prevention.* Washington, DC: National Academy Press

IOM. 1997a. *Disability Evaluation Study Design: First Interim Report.* Washington, DC: National Academy Press.

IOM. 1997b. *Enabling America: Assessing the Role of Rehabilitation Science and Engineering.* Washington, DC: National Academy Press

IOM. 1998. *The Social Security Administration's Disability Decision Process: A Framework for Research: Second Interim Report.* Washington, DC: National Academy Press.

IOM. 1999a. *Measuring Functional Capacity and Work Requirements: Summary of a Workshop.* Washington, DC: National Academy Press.

IOM. 1999b. *Review of the Disability Evaluation Study Design: Third Interim Report.* Washington, DC: National Academy Press.

IOM. 2000. *Survey Measurement of Work Disability: Summary of a Workshop.* Washington, DC: National Academy Press.

Jacobs E, ed. 1999. *Handbook of Labor Statistics.* Lanham, MD: Bernan Press.

Jette A, Assmann S, Rooks D, Harris B, Crawford S. 1998. Interrelationships among disablement concepts. *Journal of Gerontology: Med Sci* 53A(5):M395–M404.

Kennedy C, Gruenberg EM. 1987. A lexicology for the consequences of mental disorders. Chapter One in *Psychiatric Disability: Clinical, Legal and Administrative Dimensions.* Myerson AT, Fine T, eds. Washington, DC: American Psychiatric Press, Inc.

Lahiri K, Vaughan D, Wixon B. 1995. Modeling SSA's sequential disability determination process using matched SIPP data. *Social Security Bulletin* 58(4):3–42.

LaPlante MP. 1991. Disability risks of chronic illnesses and impairments. *Disability Statistics Report* (2). Washington, DC: National Institute on Disability and Rehabilitation Research, U.S. Department of Education.

LaPlante MP. 1996. *Highlights from the National Health Interview Survey Disability Study.* Presentation to the Committee to Review the Social Security Administration's Disability Decision Process Research, Institute of Medicine and Committee on National Statistics, National Research Council. Washington, DC.

LaPlante MP, Carlson D. 1996. *Disability in the United States: Prevalence and causes, 1992. Disability Statistics Report* (7). Washington, DC: U.S. Department of Education, National Institutes on Disability and Rehabilitation Research.

LaPlante MP, Kennedy J, Kaye HS, Wenger B. 1996. Disability and employment. *Disability Statistics Abstract, Number 11.* Disability Statistics Rehabilitation Research and Training Center, University of California, San Francisco.

Last JM, ed. 1983. *A Dictionary of Epidemiology.* New York: Oxford University Press.

Lawrence R, Jette A. 1996. Disentangling the disablement process. *Journal of Gerontology: Soc. Sci* 51B (4):S173–S182.

Lawton MP. 1983. Environment and other determinants of well-being in older people. *Gerontologist* 23:349–357.

Lechner D, Roth D, Straaton K. 1997. Functional capacity evaluation in work disability. *Work* 1:31–47.

Lessler JT, Kalsbeek WD. 1992. *Nonsampling Errors in Surveys.* New York: John Wiley.

Levine L. 2000. *The Employment of People with Disabilities in the 1990s.* Congressional Research Service Report for Congress. Washington, DC.

Levine S, Croog S. 1984. What constitutes quality of life? A conceptualization of the dimensions of life quality in healthy populations and patients with cardiovascular disease. In: Wenger N et al., eds. *Assessment of Quality of Life in Clinical Trials of Cardiovascular Therapies.* New York: LeJacq Publication. Co. Pp. 46–66.

Marge M. 1988. Health promotion for people with disabilities: Moving beyond rehabilitation. *American Journal of Health Promotion* 2(4):29–44.

Mashaw JL. 1983. *Bureaucratic Justice: Managing Social Security Disability Claims.* Yale University Press. New Haven and London.

Mashaw JL. 1997. Disability: Why does the search for good programs continue? In: Kingson ER, Schulz JH, eds. *Social Security in the 21st Century.* New York: Oxford University Press.

Mashaw JL, Reno VP, eds. 1996. *Disability Policy Panel Report: Balancing Security and Opportunity: The Challenge of Disability Income Policy.* National Academy of Social Insurance. Washington, DC.

Mather JH. 1993. The problem of functional assessment: political and economic perspectives. *American Journal of Occupational Therapy* 47(3):240–246.

Mathiowetz N. May 27–28, 1999. Methodological issues in the measurement of work disability. In: IOM. *Survey Measurement of Work Disability: Summary of a Workshop.* Washington, DC: National Academy Press 3:28–52.

Muller LS, Wheeler PM. 1995. Disability Program Growth: Results from Social Security's Survey of Field Office Managers. Paper presented at the joint SSA and ASPE conference, *The Social Security Administration's Disability Programs: Explanations of Recent Growth and Implications for Disability Policy.* July 1995. Washington DC.

Nagi S. 1964. A study in the evaluation of disability and rehabilitation potential: Concepts, methods, and procedures. *American Journal of Public Health* 54: 1568–1579.

Nagi S. 1965. Some conceptual issues in disability and rehabilitation. In: Sussman MB, ed. *Sociology and Rehabilitation.* Washington, DC: American Sociological Association.

Nagi S. 1976. An epidemiology of disability among adults in the United States. *Milbank Memorial Fund Quarterly: Health and Society* 54:439–467.

Nagi S. 1979. The concept and measurement of disability. Chapter One in *Disability Policies and Government Programs.* Berkowitz ED, ed. New York: Praeger Publishers.

Nagi S. 1991. Disability concepts revisited: Implications for prevention. In: IOM. *Disability in America: Toward a National Agenda for Prevention.* Washington, DC: National Academy Press.

OIG (Office of the Inspector General, SSA). 2000. *Status of the Social Security Administration's Updates to the Medical Listings.* SSA pub No A-01-99-21009. Washington DC.

Parsons T. 1951. *The Social System.* New York: The Free Press.

Parsons T. 1958. Definitions of health and illness in the light of American values and social structure. In: Jaco EG, ed. *Patients, Physicians, and Illness.* Glencoe, Ill: Free Press.

Patrick D. 1997. Rethinking prevention for people with disabilities part I: A conceptual model for promoting health. *American Journal of Health Promotion* 11(4):257–260

Patrick D, Peach H, eds. 1989. *Disablement in the Community: A Sociomedical Press Perspective.* Oxford, U.K.: Oxford University Press.

Rand M. 2001. Developing the capacity to measure crime victimization of people with disabilities. In: *Seminar on Integrating Federal Statistical Information and Processes. Statistical Policy.* Working Paper 32. Office of Management and Budget, Statistical Policy Office. Washington, DC.

Reno V. 1999. Adapting measurement of functional capacity to work to SSA's disability decision process. In: IOM. *Measuring Functional Capacity and Work Requirements. Summary of a Workshop.* Washington, DC: National Academy Press.

Rodgers W, Miller B. 1997. A comparative analysis of ADL questions in surveys of older people. *Journal of Gerontology* 52B:21-36.

Rupp K, Stapleton D. 1995. Determination of the growth in the Social Security Administration's disability programs—An overview. *Social Security Bulletin* 58(4):43–69. SSA pub No. 13–11700. Washington DC.

Sarbin T, Allen V. 1968. Role theory. In: Linsey G, Aronson E, eds. *The Handbook of Social Psychology*, Vol. 1, 2nd ed. Reading, MA: Addison-Wesley Publ. Co.

Schneider M. 2001. *Participation and Environment in the ICF and Measurement of Disability*. Paper presented at the United Nations International Seminar on the Measurement of Disability. United Nations, New York, NY.

SSA (Social Security Administration). 1981. *Annual Statistical Supplement to the Social Security Bulletin*. Washington, DC: SSA.

SSA. April 1993. (unpublished draft) *Documentation of the SSA Disability Claim and Appeal Process*. Washington, DC: SSA, Office of Human Resources, Office of Workforce Analysis.

SSA. 1994a. *Plan for a New Disability Claim Process*. Washington, DC: U.S. Government Printing Office (DHHS/SSA): SSA Pub. No. 01–005.

SSA. 1994b. *Disability Process Redesign: Next Steps in Implementation*. Washington, DC: U.S. Government Printing Office (DHHS/SSA): SSA Pub. No.01–006.

SSA. 1996. Research plan for the development of a redesigned method of evaluating disability in social security claims. *Federal Register* 61(175):47542–47544.

SSA. 1997. Progress report on development of a redesigned method of evaluating disability in social security claims. *Federal Register* 62(121):34097–34101.

SSA. 1998. *Disability Redesign—Overview and Status* [Online]. Available: http://www.ssa.gov/DPRT/DPRT_intro.html. [Accessed January 26, 1998.]

SSA. 1999a. *Annual Statistical Supplement to the Social Security Bulletin*. Washington, DC: SSA.

SSA. 1999b (unpublished draft). *Improving the Disability Adjudication Process in the Post-Redesign Era*. (Presented to the Committee to Review the Social Security Administration's Disability Decision Process Research at its October 1999 meeting.) SSA, Office of Disability.

SSA. 1999c. *Social Security Disability Insurance Program Worker Experience*. Actuarial Study No. 114. SSA Pub. No. 11-11543. Washington, DC.

SSA. 2000. *Annual Report of the Supplemental Security Income Program*. Washington, DC: SSA.

SSA. 2001a. *Disability Evaluation Under Social Security*. Washington, DC: SSA Pub. No. 64–039

SSA. 2001b. *SSI Annual Statistical Report 2000*. SSA. Washington, DC.

SSA. 2001c. *Annual Report of the Supplemental Security Income Program*. SSA. Washington, DC.

SSA. 2001d. *Annual Statistical Supplement, 2001 to the Social Security Bulletin*. SSA Publ. No. 13–11700. Washington, DC: SSA.

SSAB (Social Security Advisory Board). 1997. *Developing Social Security Policy: How the Social Security Administration Can Provide Greater Policy Leadership*. Social Security Advisory Board. Washington, DC.

SSAB. 1998. *Strengthening Social Security Research: The Responsibilities of the Social Security Administration*. Social Security Advisory Board. Washington, DC.

SSAB. 2001. *Charting the Future of Social Security's Disability Programs: The Need for Fundamental Change*. Social Security Advisory Board. Washington, DC.

SSAB. 2002. *SSA's Obligation to Ensure That the Public's Funds Are Responsibly Collected and Expended*. Social Security Advisory Board. Washington, DC.

Stapleton D, Dietrich K. 1995. *Long Term Trends and Cycles in Application and Award Growth*. Paper presented at the Conference on the Social Security Administration's Disability Programs: Explanation of Recent Growth and Implications for Disability Policy. Sponsored by Social Security Administration and the Office of the Assistant Secretary for Planning and Evaluation, DHHS, July 1995.

Stapleton D, Coleman K, Dietrich K. 1995 *Demographic and Economic Determinants of Recent Applications and Awards Growth for SSA's Disability Programs*. Paper presented at the Conference on the Social Security Administration's Disability Programs: Explanation of Recent Growth and Implications for Disability Policy. Sponsored by Social Security Administration and the Office of the Assistant Secretary for Planning and Evaluation, DHHS, July 1995.

Stapleton D, Coleman K, Dietrich K, Livermore G. 1998 Empirical analysis of DI and SSI application and award growth. In: Kalman R, Stapleton D, eds. *Growth in Disability Benefits: Explanations and Policy Implications*. Kalamazoo, Michigan: W.E. Upjohn Institute for Employment Research.

Statistics Canada. 1993. *Adults with Disabilities: Their Employment and Education Characteristics. 1991 Health and Activity Limitation Survey*. Ottawa: Statistics Canada, Catalogue 82–554.

Stoddard S, Jans L, Ripple J, Kraus L. 1998. *Chartbook on Work and Disability in the United States, 1998. An InfoUse Report*. Washington, DC: U.S. National Institute on Disability and Rehabilitation Research.

Thornberry OT, Massey JT. 1988. Trends in United States telephone coverage across time and subgroups. In Groves RM, Biemer PP, Lyberg LE, Massey JT, Nicholls WL, Waksberg J, eds. *Telephone Survey Methodology*. New York: John Wiley, 1988, pp. 25–49.

Todorov A, Kirchner C. 2000. Bias in proxies' reports of disability: Data from the National Health Interview Survey on Disability. *American Journal of Public Health*. 90(8):1248–1253.

Trupin L, Sebesta D, Yelin E, LaPlante M. 1997. Trends in labor force participation among persons with disabilities, 1983–1994. *Disability Statistics Report 10*. Washington, DC: National Institute on Disability and Rehabilitation Research, U.S. Department of Education.

U.S.House of Representatives, Committee on Ways and Means. October 2000. *2000 Green Book*. Committee Print. 106th Congress, 2nd Session. WMCP:106–14. Washington, DC.

Verbrugge LM. 1990. The iceberg of disability. In: Stahl SM, ed. *The Legacy of Longevity: Health and Health Care in Later Life*. Newbury Park, CA: Sage Publications, Inc.

Verbrugge L, Jette A. 1994. The disablement process. *Soc Sci & Med* 38(1): 1–14.

Westat. 1999a (unpublished). *Disability Evaluation Study—Instruments and Procedures: Task 4, Report 1*. Submitted to Social Security Administration, Washington, DC.

Westat. 1999b (unpublished). *Disability Evaluation Study—Final Sample Design Report: Task 4, Report 2*. (Includes Pilot Study Design.) Submitted to Social Security Administration, Washington, DC.

Westat. 1999c (unpublished). *Disability Evaluation Study—Plans to Meet Response Rate Goals: Task 4, Report 3*. Submitted to Social Security Administration, Washington, DC.

Westat. 1999d (unpublished). *Disability Evaluation Study—Pilot Study Design Report: Task 4, Report 4, Final Report*. Submitted to Social Security Administration, Washington, DC.

Whiteneck G. 2001. Validated measures of participation and the environment from Craig Hospital CHART and CHIEF. Paper presented at the International Seminar on the Measurement of the Environment held on June 4–6, 2001. United Nations. New York.

WHO (World Health Organization). 1947. *Constitution of the World Health Organization*. New York, NY: World Health Organization.

WHO. 1980. *The International Classification of Impairments, Disabilities, and Handicaps—A Manual Relating to the Consequences of Disease*. Geneva. WHO.

WHO. 1997. *ICIDH-2: International Classification of Impairments, Activities, and Participation. A Manual of Dimensions of Disablement and Functioning*. Beta-1 draft for field trials. Geneva: WHO.

WHO. 2001. *International Classification of Functioning, Disability, and Health*. Geneva: WHO.

Yelin E. 1992. *Disability and the Displaced Worker.* New Brunswick, NJ: Rutgers University Press.

Yelin E. 1999. Measuring functional capacity of persons with disabilities in light of emerging demands in the workplace. In: IOM. *Measuring Functional Capacity and Work Requirements: Summary of a Workshop.* Washington, DC: National Academy Press 2:4–27.

Yelin E, Nevitt M, Epstein W. 1980. Toward an epidemiology of work disability. *Milbank Memorial Fund Quarterly: Health and Society* 58(3):386–415

Acronyms and Abbreviations

ACS	American Community Survey
ADA	Americans with Disability Act of 1990
ADL	Activity of daily living
AHCPR	Agency for Health Care Policy and Research
AHRQ	Agency for Healthcare Research and Quality
AHS	American Housing Survey
AIR	American Institutes for Research
ALJ	Administrative law judge
AMA	American Medical Association
APA	American Psychiatric Association
APTD	Aid to the Permanently and Totally Disabled
BJS	Bureau of Justice Statistics
BLS	Bureau of Labor Statistics
BRFSS	Behavioral Risk Factors Surveillance System
CDC	Centers for Disease Control and Prevention
CDR	Continuing disability review
CDSR	Clinical Disability Severity Rating
CNSTAT	Committee on National Statistics (NRC)
CPS	Current Population Survey
CWHS	Continuous Work History Sample

DBSSE	Division on Behavioral and Social Sciences and Education (NRC)
DDS	Disability Determination Service
DES	Disability Evaluation Study
DHHS	U.S. Department of Health and Human Services
DO	District office (SSA)
DOL	U.S. Department of Labor
DOT	Dictionary of Occupational Titles
DRI	Disability Research Institute
DSM-III	*Diagnostic and Statistical Manual of Mental Disorders*, Third Edition
GAO	U.S. General Accounting Office
HALS	Health and Activity Limitation Survey
HRS	Health and Retirement Survey
IADL	Instrumental activity of daily living
ICD	International Classification of Diseases and Related Health Problems
ICF	International Classification of Functioning, Disability, and Health
ICIDH	International Classification of Impairments, Activities, and Participation
IOM	Institute of Medicine
IRS	Internal Revenue Service
MBR	Master Beneficiary Record
MEC	Mobile examination center
MEF	Master Earnings File
MEPS	Medical Expenditure Panel Survey
MH/ABCD Task Force	Task Force on Mental Health, and Addictive, Behavioral, Cognitive, and Developmental Aspects of ICIDH
MRFCA	Mental Residual Functional Capacity Assessment
NACC	North American Collaborating Center (ICF)
NAS	National Academy of Sciences
NASI	National Academy of Social Insurance
NCHS	National Center for Health Statistics
NCVS	National Crime Victimization Survey
NHANES	National Health and Nutrition Examination Survey
NHIS	National Health Interview Survey
NHIS-D	National Health Interview Survey on Disability

NHSDA	National Household Survey on Drug Abuse
NLS	National Longitudinal Study
NRC	National Research Council
NSFG	National Survey of Family Growth
NSHA	National Study of Health and Activity

OAA	Old Age Assistance
OASDI	Old Age, Survivors, and Disability Insurance
OD	Office of Disability (SSA)
OIG	Office of the Inspector General
O*NET	Occupational Information Network
ORES	Office of Research, Evaluation and Statistics (SSA)

| **PRTF** | Psychiatric Review Technique Form |
| **PSU** | Primary sampling unit |

RDD	Random digit dialing
RFC	Residual functional capacity
RRC	Retirement Research Consortium
RTI	Research Triangle Institute

SGA	Substantial gainful activity
SIPP	Survey of Income and Program Participation
SSA	Social Security Administration
SSAB	Social Security Advisory Board
SSDI	Social Security Disability Insurance
SSI	Supplemental Security Income
SSR	Supplemental Security Record

| **VCU** | Virginia Commonwealth University |

| **WHO** | World Health Organization |
| **WHO DAS** | WHO Disability Assessment Schedule |

Appendix A

Committee Meetings and Presenters of Testimony

The Committee to Review the Social Security Administration's Disability Decision Process Research held a total of 12 meetings starting in 1997. These meetings involved segments open to the public, as well as closed sessions for committee deliberation. The dates of these meetings are listed below:

January 27–28, 1997, Washington, D.C.
May 30–31, 1997, Washington, D.C.
October 17–18, 1997, Washington, D.C.
March 6–7, 1998, Washington, D.C.
October 8–9, 1998, Washington, D.C.
March 29–30, 1999, Washington, D.C.
September 30–October 1, 1999, Washington, D.C.
April 13–14, 2000, Washington, D.C.
September 15, 2000, Washington, D.C.
April 5–6, 2001, Washington, D.C.
September 6–7, 2001, Washington, D.C.
December 20–21, 2001, Washington, D.C.

The committee heard from numerous presenters at these meetings. They include

Alexander Vachon, U.S. Congress
David Podoff, U.S. Congress

Kim Hildred, U.S. Congress
Sandy Wise, U.S. Congress
John Kregel, Virginia Commonwealth University
Donna Dye, Department of Labor
Teresa Russell, American Institutes for Research
Gary Kay, Consultant
Dixie Sommers, National O*NET Consortium
Randall Keesling, National O*NET Consortium
Phill Lewis, National O*NET Consortium
David Osborne, Library of Congress
Gerry Hendershot, NCHS
Cynthia Thomas, Westat
Mitchell LaPlante, University of California at San Francisco
Susan Van Hemel, DBSSE
Christine R. Hartel, DBSSE

Appendix B

Workshop Agendas and Presenters

Workshop on Functional Capacity and Work Requirements
as It Relates to SSA's Disability Decision Process Research

June 4–5, 1998

Committee to Review SSA's Disability Decision Process Research
National Academy of Sciences/Institute of Medicine
Cecil and Ida Green Building, Conference Room 104
2001 Wisconsin Avenue, N.W., Washington, D.C.

FINAL AGENDA

Thursday, June 4

8:30–9:00 a.m.   *Continental Breakfast*

9:00–9:15 a.m.   Welcome and Introduction
                    Dorothy Rice, *Chair*

9:15–9:25 a.m.   Purpose and Goals of the Workshop
                    Dorothy Rice

OPENING SESSION

9:25–10:30 a.m.   **Measuring Functional Capacity of Persons with Disabilities in Light of Emerging Demands in the Workplace**
(Commentary and discussion will follow)
Paper Presented By:
Edward Yelin
Discussant:
Janet Norwood

10:30–10:45 a.m.   *Coffee Break*

SESSION ONE

10:45–12:30 p.m.   **Linking Components of Functional Capacity Domains (Cognitive, Psychosocial, Motor and Sensory/Perceptual) with Work Requirements**
• What are the specific components of the functional capacity domains?
• How are the specific components linked to demands of work?
• Is it possible to develop a baseline of work requirements? Can O*NET be used or adapted to meet SSA's need for an occupational classification system?
Discussion Leader:
Howard Goldman
Discussants:
Edwin Fleishman
Cille Kennedy

12:30–1:30 p.m.   *Lunch in Refectory*

SESSION TWO

1:30–3:30 p.m.   **Desired Characteristics of Instruments to Measure Functional Capacity to Work**
• What are the strengths and limitations of self-reports, proxy reports, performance testing, and clinical observation?

- How do the strengths and weaknesses of different measurement approaches vary across the different domains of functioning?
- To what extent should assistive devices be considered in measuring functional capacity?
- Do different populations have different measurement requirements (e.g., schizophrenia vs. arthritis vs. spinal injury vs. Alzheimer's disease)?

Discussion Leader:
Alan Jette
Discussants:
Allen Heinemann
Constantine Lyketsos

3:30–3:45 p.m.   *Coffee Break*

SESSION THREE

3:45–4:45 p.m.   **The Use of Functional Capacity Measures in Public and Private Programs in the United States and in Other Countries**

- What has been their experience in the use of functional capacity measures in determining disability?
- What aspects of their measurement of functional capacity might be relevant for SSA's needs?

Discussion Leader:
Patricia Owens
Discussants:
Richard Burkhauser
Ian Basnett

4:45–5:30 p.m.   General Discussion

5:30 p.m.   Adjourn—*Reception*

6:30 p.m.   *Dinner for Committee Members and Invited Guests*

Friday, June 5

SESSION FOUR

8:30–9:00 a.m.   *Continental Breakfast*

9:00–10:15 a.m.   **Adapting Measurement of Functional Capacity
                      to Work to SSA's Disability Decision Process**
                      • What are the criteria for a "successful" measure-
                        ment of functional capacity to work?
                      • Feasibility and practicality of designing and ad-
                        ministering (i.e., safety, cost, etc.) measures of
                        functional capacity to work.
                      • Technical issues of incorporating reliability, va-
                        lidity, sensitivity, and specificity in the context
                        of SSA's disability decision process.
                      • How can these measurement approaches be
                        linked to work requirements in the context of
                        SSA's disability decision process?
                      Discussion Leader:
                      Virginia Reno
                      Discussants:
                      Lisa Iezzoni
                      David Stapleton

10:15–10:30 a.m.   *Coffee Break*

10:30–11:00 a.m.   **Rapporteur's Review of Major Issues Identified**
                      Jane West
                      Kristen Robinson

11:00–12:00 p.m.   General Discussion

12:00–12:15 p.m.   Concluding Remarks
                      Dorothy Rice

12:15 p.m.   Adjourn

Workshop on Survey Measurement of Work Disability:
Challenges for Survey Design and Method

May 27–28, 1999

Committee to Review SSA's Disability Decision Process Research
National Academy of Sciences/Institute of Medicine
Holiday Inn-Georgetown, Mirage I
2101 Wisconsin Avenue, N.W., Washington, D.C.

AGENDA

Thursday, May 27

8:30–9:00 a.m.   *Continental Breakfast*

9:00–9:15 a.m.   Welcome and Introduction
                   Dorothy Rice, *Chair*

9:15–9:30 a.m.   Welcoming Remarks
                   Jane Ross, *Deputy Commissioner, SSA*

SESSION ONE

9:30–10:30 a.m.   **Overview of the Two Background Papers:
                   Opportunities for Methodological Research on
                   Survey Measures Related to Disability**
                   An examination of the various conceptual models
                   of disability and the disablement process and their
                   ability to address SSA's disability program require-
                   ments.
                   • The challenges related to the translation of con-
                     ceptual models to valid and reliable questions
                     that can be administered to the general popula-
                     tion.
                   • The identification of the coverage, nonresponse,
                     and measurement error properties of current
                     measures of work disability.
                   • Potential problems in cross-walking among
                     measures of disability collected in a variety of
                     settings and under varying survey conditions.

Elizabeth Badley
Alan Jette
Nancy Mathiowetz
Contributors:
Allan Sampson

10:30–10:45 a.m.   *Coffee Break*

SESSION TWO

10:45 a.m.–   **Implications of Different Concepts for Survey**
12:00 noon   **Measurement Problems**
                     • How do the various conceptual models address
                       the dynamic nature of disability and how do
                       these models address SSA's disability program
                       requirements?
                     • How do the various conceptual models address
                       the role of environment, adaptation, expecta-
                       tions, and perceptions?
                     • What measurement gaps exist between the vari-
                       ous conceptual models of disability and the cur-
                       rent set of disability measures used in federal
                       surveys?
                     Discussion Leader:
                     Robert Groves
                     Contributors:
                     Ellen MacKenzie
                     Allan Hunt

12:00 noon–1:00 p.m. *Lunch in Kaleidoscope Room*
                      (Committee members and invited guests)

SESSION THREE

1:00–2:00 p.m.   **Sampling, Accessing, and Measuring People**
                      with Disabilities
                      • To what extent do varying modes and methods
                        of data collection facilitate participation
                        among persons with disabilities?
                      • If access to a person with a work disability is
                        limited (due to the interface between the survey
                        design and the nature of the disability), how is
                        the measurement of disability affected by the role

of the proxy respondent—caregiver as respondent, other proxy respondent? Can trade-offs be assessed between nonresponse and measurement errors?
• What gaps exist in our knowledge of the relative impact of coverage, nonresponse, and measurement error on estimates of disability?
Discussion Leader:
Colm O'Muircheartaigh
Contributors:
Lawrence Branch
Ronald Kessler

SESSION FOUR

2:00–3:00 p.m.   **Questionnaire Development Issues for Measures of Work Disability**
• In light of developments related to the integration of cognitive theory and survey methodology, how should measures of work disability be evaluated?
• How does the dynamic nature of disability and the disablement process impact the measurement of work disability?
• How is measurement affected by the role of the person providing the information—self-respondent, caregiver as respondent, or other proxy reporters?
• To what extent should we look to statistical modeling related to scale reduction as a means for reducing the effects of measurement error?
• How will the measurement of work disability in a variety of settings (the DES and other ongoing federal data collection efforts) impact SSA's ability to monitor the pool of people potentially eligible for disability benefits?
• What research needs to be conducted to develop robust measures of work disability and to address the gaps in our knowledge about the measurement error properties of current measures?
Discussion Leader:
Seymour Sudman

Contributors:
Roger Tourangeau
Jack McNeil

3:00–3:30 p.m.   *Coffee Break*

SESSION FIVE

3:30–5:00 p.m.   **Role of Environment in Survey Measurement of Disability**
- How is the measurement of work disability affected by environment, perceptions, and expectations?
- Is there a differential impact of environment on the reporting of disability as a function of the role of the person providing the information—self-respondent, caregiver, or other proxy respondent?
- What do we know about the measurement of the role of environment, expectations, and perceptions with respect to the various sources of survey error, specifically, nonresponse and measurement error?
- What gaps exist in our knowledge of how to adequately measure environment and its impact on the measurement of work disability? What research needs to be conducted to address these gaps?

Discussion Leader:
David Gray
Contributors:
Sandra Berry
Lois Verbrugge

5:00–5:30 p.m.   General Discussion

5:40–6:40 p.m.   Adjourn—*Reception for all attendees*

6:45 p.m.   *Dinner in Kaleidoscope Room*
                        (Committee members and invited guests)

Friday, May 28

SESSION SIX

8:30–9:00 a.m.	*Continental Breakfast*

9:00–10:30 a.m.   **Defining a Research Agenda**
- What are the criteria for a "successful" measurement of functional capacity to work?
- Feasibility and practicality of designing and administering (i.e., safety, cost, etc.) measures of functional capacity to work.
- Technical issues of incorporating reliability, validity, sensitivity, and specificity in the context of SSA's disability decision process.
- How can these measurement approaches be linked to work requirements in the context of SSA's disability decision process?

    Discussion Chair:
    Dorothy Rice

10:30–10:45 a.m.   *Break*

10:45 a.m.–   General Discussion
12:00 noon   *Moderator:*
    Robert Groves

12:00–12:15 p.m   Concluding Remarks
.   Dorothy Rice

12:15 p.m.   Adjourn

Appendix C

Committee Recommendations

The following is a compilation of all recommendations made by the committee in the interim reports.

DISABILITY EVALUATION STUDY DESIGN

First Interim Report

RECOMMENDATION 3-1. The committee strongly endorses the conduct by the Social Security Administration of a well-designed, carefully pretested, and statistically sound Disability Evaluation Study.

RECOMMENDATION 3-2. The committee recommends that the current stage 1 and pilot study be merged, expanded, and extended into a research, development, and testing phase of the survey with application to samples of the type that are more traditionally used in methods testing. Only when the development and refinement of the functional assessment instruments, survey operations, and other issues are tested and resolved should a national sample survey be launched using a single protocol.

RECOMMENDATION 3-3. The committee recommends that the national survey should be conducted with one sample large enough to estimate the sizes of the populations at risk with acceptable levels of statistical precision.

RECOMMENDATION 3-4. The committee recommends that the Social Security Administration use relevant data from the National Health Interview Survey Disability Supplement, National Health and Nutrition Examination Survey, Survey of Income and Program Participation, and other relevant surveys to assist in developing the sample design, survey operation, and questionnaire content for the Disability Evaluation Study.

RECOMMENDATION 4-5. The committee recommends that the Disability Evaluation Study be based on a design offering full coverage of the U.S. household population of adults. If resources are lacking to mount an area probability sample using face-to-face interviews, the Social Security Administration should use a multiple frame design of a statistically optimum mix of the general population followed by face-to-face interviews of the eligible population.

RECOMMENDATION 4-6. The committee recommends that once the options for using different combinations of team composition and origin, examination setting, and other dimensions are sufficiently set for assessments, a formal field experiment should be performed during the research, development, and testing phase of the survey to determine the validity and reproducibility of these options as well as the most cost-effective approach to meeting the objectives of the survey.

RECOMMENDATION 4-7. The committee recommends that the Social Security Administration require in the scope of work a rigorously designed experiment in the field testing and development phase of the survey to identify mechanisms for enhancing participation in the Disability Evaluation Study, to guide decisions on the use of home examination for those unable to travel to an examination site, to establish the validity of the measures obtained, and to assess the quality of the medical evidence of record.

RECOMMENDATION 4-8. The committee recommends that the Social Security Administration enhance the safeguards of matched data according to accepted practices by employing procedures used in recent federal surveys and that it take into consideration the effect of such procedures on response rates.

THE SOCIAL SECURITY ADMINISTRATION'S DISABILITY DECISION PROCESS: A FRAMEWORK FOR RESEARCH

Second Interim Report

RECOMMENDATION 4-1. The committee recommends that early in the redesign effort, the Social Security Administration should specify how it will define, measure, and assess the criteria it will use to evaluate the current disability determination process, as well as any alternative processes being developed.

RECOMMENDATION 4-2. The committee recommends that the Social Security Administration develop an alternative plan for use of functional assessment measures in the disability decision process in the event that the proposed global, standardized, functional assessment measure is not developed and tested in time for implementation.

RECOMMENDATION 4-3. The committee recommends that the Social Security Administration develop an interim plan for an occupational classification system in the event that the Occupational Information Network (O*NET) database is either not completed or insufficient to meet the needs of a new disability decision process.

RECOMMENDATION 4-4. The committee recommends that the Social Security Administration conduct baseline studies on the role of the evaluation of vocational factors in the current decision-making process and the effects of these factors on the populations of claimants and beneficiaries.

RECOMMENDATION 4-5. The committee recommends that the Social Security Administration reconsider the timeframe for completion of the redesign research so that the necessary questions can be answered in an appropriately sequenced and coordinated manner.

RECOMMENDATION 4-6. The committee recommends that the Social Security Administration establish a cognitive laboratory for the Disability Evaluation Study, disability decision process research, and for other purposes of the agency.

RECOMMENDATION 4-7. The committee recommends that the Social Security Administration actively engage process engineering experts (such as industrial engineers, operations researchers) to evaluate and improve the Social Security Administration's disability benefits administrative process to assure that task assignments and participant roles achieve a maximum level of effectiveness and efficiency.

RECOMMENDATION 4–8. The committee recommends that the Social Security Administration develop plans for simulation and modeling of alternative disability decision processes and other policy options, and devote adequate resources for this activity.

RECOMMENDATION 5–1. The committee recommends that the Social Security Administration's research and evaluation staff and its extramural research program be expanded substantially.

REVIEW OF THE DISABILITY EVALUATION STUDY DESIGN

Third Interim Report

RECOMMENDATION: The committee strongly recommends that SSA revise the project schedule to allow significantly more time to plan and analyze the pilot study and test alternative solutions for problem areas before starting the national study.

Biographical Sketches of
Committee Members

Dorothy P. Rice, Sc.D. (Hon.) (*Chair*) is Professor Emeritus of Medical Economics at the Institute for Health and Aging, School of Nursing, University of California at San Francisco (UCSF). From 1983 to 1994, she was Professor-in-Residence at UCSF. Previously she served as Director of the National Center for Health Statistics and was Deputy Assistant Commissioner for Research and Statistics at the Social Security Administration. Professor Rice's major research interests and expertise include health statistics; survey research, design, and methods; disability; chronic illness; cost-of-illness studies; and the economics of medical care. She has achieved national and international renown for her leadership role, extensive research, and scholarly publications. Professor Rice has received numerous awards including an honorary Doctor of Science from the College of Medicine and Dentistry of New Jersey. She is a Fellow of the American Public Health Association and the American Statistical Association, and a member of the Institute of Medicine.

Monroe Berkowitz, Ph.D., is Professor Emeritus of Economics and Director of Disability and Health Economics in the Bureau of Economic Research at Rutgers University. He has served as a consultant to various government agencies including the Social Security Administration, the World Health Organization, and the American Association for the Advancement of Science. Dr. Berkowitz is a leading authority on the economics of disability and rehabilitation in public programs (SSA disability insurance and workers' compensation), private disability insurance, and public and private rehabilitation systems; and has conducted extensive comparative analysis of foreign systems. He is a member of the National Academy of Arbitrators, the National Academy of Social Insurance, the American Economic Association, and the Industrial Relations Research Association.

Ronald S. Brookmeyer, Ph.D., is Professor of Biostatistics and Epidemiology at the Johns Hopkins University School of Hygiene and Public Health. He has been a Visiting Biostatistician at the National Cancer Institute and the International Agency for Research on Cancer in Lyon, France. Dr. Brookmeyer's research interests and expertise are in statistical modeling and methodology, biometrics, and epidemiology. He is the recipient of the Spiegelman Gold Medal awarded by the American Public Health Association for contributions to health statistics. He is a Fellow of the American Statistical Association and the American Association for the Advancement of Science, and a member of the Biometrics Society and the Society for Epidemiological Research.

Marshal F. Folstein, M.D., is Chair and Professor of Psychiatry at Tufts University School of Medicine and Psychiatrist-in-Chief at the New England Medical Center (NEMC). Prior to joining NEMC, he was Eugene Meyer III Professor of Psychiatry and Medicine at the Johns Hopkins Medical Institutions. His expertise and research interests are in neuropsychiatry, disability research, and Alzheimer's disease. Dr. Folstein created the Mini-Mental State Examination, widely used for assessing cognitive mental status in medical patients and in population surveys. He is a Fellow of the American College of Physicians, the American Psychiatric Association, and the Gerontological Society; and a member of the American Neurological Association and the Society for Epidemiological Research.

Robert M. Groves, Ph.D., is a Professor of Sociology and Senior Research Scientist at the University of Michigan, and a research professor at the Joint Program in Survey Methodology, based at the University of Maryland, a consortium of the University of Maryland, University of Michigan, and Westat, Inc. He is Director of the University of Michigan Survey Research Center. From 1990 to 1992, Dr. Groves was an Associate Director of the U.S. Census Bureau, on loan from Michigan. He has over 25 years of experience with large-scale surveys, and has investigated the impact of alternative telephone sample designs on precision, the effect of data collection mode on the quality of survey reports, causes and remedies for nonresponse errors in surveys, estimation and explanation of interviewer variance in survey responses, and other topics in survey methods. His current research interests focus on theory-building in survey participation and models of nonresponse reduction and adjustment. He is a Fellow of the American Statistical Association, an elected member of the International Statistical Institute, former President of the American Association for Public Opinion Research, and former Chair of the Survey Research Methods Section of the American Statistical Association.

Alan M. Jette, Ph.D., is Professor and Dean of Boston University's Sargent College of Health and Rehabilitation Sciences, and Professor of Social and Behavioral Sciences at the Boston University School of Public Health. His previous appointments have included Chief Research Scientist, New England Research Institute; Associate Professor, Massachusetts General's Institute of Health Professions; and Assistant Professor, Division on Aging, Harvard Medical School. Dr. Jette currently directs the Edward R. Roybal Research Center on Enhancing Late-Life Function, funded by the National Institute on Aging. Within the Roybal Center, he and his colleagues are testing physical activity and other intervention strategies designed to prevent late-life disability. He also directs Boston University's Rehabilitation Research and Training Center on Measuring Research. In this Center, he and his colleagues are applying modern psychometric methods to the development of the next generation of outcome instruments for use in rehabilitation.

William D. Kalsbeek, Ph.D., is Professor of Biostatistics and Director of the Survey Research Unit at the University of North Carolina-Chapel Hill. His prior experience includes statistical research with the Office of Research and Methodology at the National Center for Health Statistics and at the Sampling Research and Design Center at the Research Triangle Institute in North Carolina. Dr. Kalsbeek's research interests and areas of expertise are in biostatistics, survey design and research, and assessment; and he is well known for his work in survey methods. He is a Fellow of the American Statistical Association and a member of the American Public Health Association.

Jerry L. Mashaw, LL.B., Ph.D., is Sterling Professor of Law and Management and Professor at the Institute of Social and Policy Studies at Yale University. He is a leading scholar in administrative law and has written widely on social insurance, social welfare issues, and disability policy. Dr. Mashaw recently chaired the National Academy of Social Insurance's Disability Policy Panel. He is a Fellow of the National Academy of Arts and Sciences and founding coeditor of the *Journal of Law Economics and Organization*.

Catharine C. (Katie) Maslow, M.S.W., is Director of the Initiatives on Managed Care and Acute Care at the Alzheimer's Association. Prior to this, she was at the U.S. Office of Technology Assessment (OTA), and has experience in public welfare, mental health, and nursing home settings. Her research and consumer interests include aging, disability, criteria for long-term care, client assessment, and Alzheimer's disease. Ms. Maslow is a member of the National Association of Social Workers, the American Public Health Association, the Gerontological Society of America, and the American Society on Aging.

Donald L. Patrick, Ph.D., M.S.P.H., is Professor of Health Services and Director of the Social and Behavioral Sciences Program at the University of Washington School of Public Health. He holds adjunct appointments in epidemiology, sociology, pharmacy, and rehabilitation medicine, and is a senior investigator at the University's Center for Disability Policy and Research. Dr. Patrick's research interests and expertise are in social determinants of health, adolescent health, health policy for people with disabilities, and health and quality of life assessment. He is a Fellow of the Association of Health Services Research, and a member of the American Public Health Association, the British Society of Social Medicine, and the Society for Disability Studies. He was the inaugural president of the International Society for Quality of Life Research and is a member of the Institute of Medicine.

Harold A. Pincus, M.D., is the Executive Vice Chairman of the Department of Psychiatry at the University of Pittsburgh School of Medicine and Western Psychiatric Institute and Clinics. He is also a Senior Scientist at RAND and directs the RAND Health Institute in Pittsburgh. Dr. Pincus directs the Robert Wood Johnson Foundation's National Program on Depression in Primary Care: Linking Clinical and Systems Strategies. He was the Deputy Medical Director of the American Psychiatric Association (APA) and founding director of the APA's Office of Research, and Executive Director of the American Psychiatric Institute for Research and Education. Previously, Dr. Pincus was the Special Assistant to the Director of the National Institute of Mental Health. Dr. Pincus received the William C. Menninger Memorial Award of the American College of Physicians for distinguished contributions to the science of mental health and the Health Services Research Senior Scholar Award of the American Psychiatric Association for outstanding contributions to the field. He also maintains a small private practice specializing in major affective disorders and has worked for 22 years at a public mental health clinic, caring for patients with severe mental illnesses. Dr. Pincus has led major health policy research and training projects on the interrelationships among general medical care, mental health, and substance abuse; diagnosis, classification, and treatment of mental disorders; research career development; and assessment of disability and functioning.

Edward H. Yelin, Ph.D., is Professor of Medicine and Health Policy at the University of California, San Francisco, where he has primary academic appointments in the Department of Medicine and Institute for Health Policy Studies. He is also the Director of the Arthritis Research Group at UCSF. Dr. Yelin's research interests concern the impact of managed care on persons with chronic conditions and disability and employment problems among persons with disabilities. He has over 110 publications in these areas, including *Disability and the Displaced Worker* (Rutgers University Press). Dr. Yelin is a member of the American Public Health Association and American College of Rheumatology. He has received many academic awards, including the Distinguished Scholar Award from the Association of Rheumatology Health Professionals. He was recently elected to membership in the National Academy of Social Insurance.

Part II

COMMISSIONED PAPERS

PART II

Commissioned Papers

The committee commissioned five background papers from experts in areas of survey design and method, concept and measurement of disability, mental impairments, and disability and the labor market:

1. "Conceptual Issues in the Measurement of Work Disability," by Alan Jette, Ph.D., and Elizabeth Badley, M.D. (2000)
2. "Methodological Issues in the Measurement of Work Disability," by Nancy Mathiowetz, Ph.D. (2000)
3. "SSA's Disability Determination of Mental Impairments: A Review Toward an Agenda for Research," by Cille Kennedy, Ph.D. (2001)
4. "Survey Design Options for the Measurement of Persons with Work Disabilities," by Nancy Mathiowetz, Ph.D. (2001)
5. "Persons with Disabilities and Demands of the Contemporary Labor Market," by Edward Yelin, Ph.D., and Laura Trupin, MPH (2001)

Each paper can be found in its entirety beginning on the following page.

Conceptual Issues in the Measurement of Work Disability[1]

Alan M. Jette, Ph.D., and Elizabeth Badley, M.D.[2]

The field of disability research is in need of uniform concepts and a common language to guide scholarly discussion, to advance theoretical work on the disablement process, to facilitate future survey and epidemiological research, and to enhance understanding of disability on the part of professionals as well as the general public. A commonly understood language can also influence the development of public policy in the area of work disability, the focus of the Institute of Medicine's workshop titled "Survey Measurement of Work Disability." The current lack of a uniform language and commonly understood definition of the concepts of "disability" and "work disability" is a serious obstacle to all these endeavors.

Conceptual confusion is a particular barrier to the improvement of the Social Security Administration's (SSA) process for determining eligibility for both Social Security Disability Insurance (SSDI) and Supplemental Security Income (SSI) related to "work disability," as was illustrated in the earlier Institute of Medicine workshop, "Measuring Functional Capacity and Work Requirements." A shared language and conceptual understand-

[1]This paper was originally prepared for the committee workshop titled "Workshop on Survey Measurement of Work Disability: Challenges for Survey Design and Method" held on May 27–28, 1999, in Washington, D.C. (IOM, 2000).

[2]Alan Jette is a Professor and Dean of the Sargent College of Health and Rehabilitation Sciences at Boston University. Elizabeth Badley is Director of the Arthritis Community Research & Evaluation Unit at the University Health Network in Toronto, Ontario.

ing did not emerge from that workshop. If various participants in the disability benefit determination revision process cannot agree on the meaning of the term "work disability," they can hardly be expected to reach agreement on an approach to improving the work disability determination process.

The Social Security Act defines disability as the "inability to engage in any substantial gainful activity by reason of a medically determinable physical or mental impairment which can be expected to result in death or can be expected to last for a continuous period of not less than 12 months." As this background paper will illustrate, this definition in the Social Security Act is at odds with most contemporary thought about the concept of disability and is in itself a barrier to the SSA's work disability revision process.

The paper aims to provide the reader with a conceptual foundation to facilitate discussion at the upcoming workshop titled "Survey Measurement of Work Disability." Our intent is to highlight issues regarding language and concepts directly or indirectly related to the concept of "work disability." To do so, we focus on several activities:

1. present a review of some of the contemporary definitions of disability found in the literature;
2. discuss these definitions in the context of several major disablement frameworks;
3. discuss the concept of "work disability" in the context of these disablement models and relate it to other health-related phenomena;
4. critically review the conceptual basis of frequently used survey items that attempt to assess "work disability"; and
5. highlight some of the pressing research needs in the area of "work disability."

THE CONCEPT OF DISABILITY

A common understanding of the term "disability" is an essential first step to a scholarly exchange about the concept of "work disability" and is the foundation for a fruitful discussion of improving survey research in the general area of disability and, more specifically, in the area of work disability.

Understanding of the source of contemporary conceptual confusion requires a review of the major disability frameworks found in the literature. The goal of bringing together the several different schools of thought on disability and the disablement process remains elusive. Achieving a commonly accepted conceptual language is one of the primary challenges facing the field of disability research.

Major Schools of Thought

Several schools of thought have defined disability and related concepts. We will focus on the Disablement Model developed by Nagi (1965) and the International Classification of Impairments, Disabilities, and Handicaps (ICIDH-1) (WHO, 1980) and the current proposals for its revision, which is referred to in this paper as ICIDH-2 (WHO, 1997). We will briefly review both of these conceptual frameworks. Both the Nagi Disablement Model and the ICIDH frameworks have in common the view that overall disablement represents a series of related concepts that describe the consequences or impact of a health condition on a person's body, on a person's activities, and on the wider participation of that person in society. In the authors' view, the major differences in these frameworks are in the terms used to describe disability and related concepts and the placement of the boundaries between concepts more than differences in their fundamental contents. After reviewing the terms within each framework we will compare and contrast the two major models along with their major derivatives and explore how these relate more generally to the concept of "work disability."

Nagi's Concept of Disability

According to the conceptual framework of disability developed by sociologist Saad Nagi (1965), *"disability is the expression of a physical or a mental limitation in a social context."* In striking contrast to the Social Security Act's definition of work disability as an inability to work due to a physical or mental impairment, Nagi specifically views the concept of disability as representing the gap between a person's capabilities and the demands created by the social and physical environments (Nagi, 1965, 1976, 1991). This is a fundamental distinction of critical importance to scholarly discussion and research related to disability phenomena.

According to Nagi's own words:

[Disability is a] limitation in performing socially defined roles and tasks expected of an individual within a sociocultural and physical environment. These roles and tasks are organized in spheres of life activities such as those of the family or other interpersonal relations; work, employment, and other economic pursuits; and education, recreation, and self-care. Not all impairments or functional limitations precipitate disability, and similar patterns of disability may result from different types of impairments and limitations in function. Furthermore, identical types of impairments and similar functional limitations may result in different patterns of disability. Several other factors contribute to shaping the dimensions and severity of disability. These include (a) the individual's definition of the situation and reactions, which at times compound the limitations; (b) the definition of the situation by others, and their reactions and expectations—especially those who are significant in the lives of the person with the disabling condition (e.g., family members, friends and associates, employers and co-workers, and organizations and professions that provide services and benefits); and (c) characteristics of the environment and the degree to which it is free from, or encumbered with, physical and sociocultural barriers. (Nagi, 1991, p. 315)

Nagi's definition stipulates that a disability may or may not result from the interaction of an individual's physical or mental limitations with the social and physical factors in the individual's environment. Consistent with Nagi's concept of disability, an individual's physical and mental limitations would not invariably lead to work disability. Not all physical or mental conditions would precipitate a work disability, and similar patterns of work disability may result from different types of health conditions. Furthermore, identical physical and mental limitations may result in different patterns of work disability.

Nagi's Disablement Model has its origins in the early 1960s. As part of a study of decision making in the SSDI program, Nagi (1964) constructed a framework that differentiated disability (as defined and discussed above) from three other distinct yet interrelated concepts: active pathology, impairment, and functional limitation. This conceptual framework has come to be referred to as Nagi's Disablement Model.

For Nagi, *active pathology* involves the interruption of normal cellular processes and the simultaneous homeostatic efforts of the organism to regain a normal state. He notes that active pathology can result from infection, trauma, metabolic imbalance, degenerative disease processes, or other etiology. Examples of active pathology are the cellular disturbances consistent with the onset of disease processes such as osteoarthritis, cardiomyopathy, and cerebrovascular accidents.

For Nagi, *impairment* refers to a loss or abnormality at the tissue, organ, and body system level. Active pathology usually results in some type of impairment, but not all impairments are associated with active pathology (e.g., congenital loss or residual impairments resulting from trauma). Impairments can occur in the primary locale of the underlying pathology (e.g., muscle weakness around an osteoarthritic knee joint), but they may also occur in secondary locales (e.g., cardiopulmonary deconditioning secondary to inactivity).

To describe the distinct consequences of pathology at the level of the individual, Nagi uses the term *functional limitations* to represent restrictions in the basic performance of the person. An example of basic functional limitations that might result from a cerebrovascular accident could include limitations in the performance of locomotor tasks, such as the person's gait, and basic mobility, such as transfers, or in nonphysical tasks, such as communication or reasoning. Such functional limitations might or might not be related to specific impairments (secondary to the cerebrovascular accident) and thus are seen as distinct from organ or body system disturbances.

At this point, a "work disability" example will illustrate the distinctions being drawn between the various concepts within Nagi's Disablement Model. Two patients with Parkinson's disease may enter the Social Security work disability benefits determination process with very similar clinical profiles. Both may have moderate impairments such as rigidity and bradykinesia. Their patterns of function may also be similar with a characteristically slow, shuffling gait, and slow deliberate movement patterns. Their work role patterns, however, may be radically different. One individual may have restricted his or her outside activities completely, need help dressing in the morning, spend most of the time indoors watching television, be depressed, and be currently unemployed. The other may be fully engaged in his or her social life, receive assistance from a spouse in performing daily activities, be driven to work, and, through workplace modification, be able to maintain full-time employment. The two patients present very different work disability profiles yet have very similar underlying pathology, impairment, and functional limitation profiles.

Elaboration of Nagi's Disablement Model

In their work on the disablement process, Verbrugge and Jette (1994) maintained the basic concepts of the Nagi Disablement Model and Nagi's original definitions. Within the dimension of disability, however, they categorized subdimensions of social roles that can be considered under Nagi's concept of disability. Some of the most commonly applied dimensions include the following:

- *Activities of daily living (ADL)*—including behaviors such as basic mobility and personal care.
- *Instrumental activities of daily living (IADL)*—including activities such as preparing meals, doing housework, managing finances, using the telephone, and shopping.
- *Paid and unpaid role activities*—including performing one's occupation, parenting, grandparenting, and being a student.

- *Social activities*—including attending church and other group activities and socializing with friends and relatives.
- *Leisure activities*—including participating in sport and physical recreation, reading, or taking distant trips.

Within their framework, "work disability" is clearly delineated as a specific subdimension under the concept of disability.

In their 1994 work, Verbrugge and Jette attempted to extend Nagi's Disablement Model to attain full sociomedical scope. They attempted to clearly differentiate the "main pathways" of the disablement process (i.e., Nagi's original concepts) from factors hypothesized or known to influence the ongoing process of disablement (Figure 1).

Viewed from a social epidemiological perspective, Verbrugge and Jette (1994) argued that one might analyze differences in disablement concepts relative to three sets of variables: predisposing risk factors, intraindividual factors, and extraindividual factors. These categories of variables, which are external to the main disablement pathway, can be defined as follows:

- *Risk factors* are predisposing phenomena that are present before the onset of the disabling event and that can affect the presence or severity of the disablement process. Examples include sociodemographic background, lifestyle, and biological factors.
- The next class of variables is *intraindividual factors* (those that operate within a person), such as lifestyle and behavioral changes, psychosocial attributes and coping skills, and activity accommodations made by the individual after the onset of a disabling condition.
- *Extraindividual factors* (those that perform outside or external to the person) pertain to the physical as well as the social context in which the disablement process occurs. Environmental factors relate to the social as well as the physical environmental factors that bear on the disablement process. These can include medical and rehabilitation services, medications and other therapeutic regimens, external supports available in the person's social network, and the physical environment.

A further elaboration of Nagi's conceptual view of the term disability is contained in *Disability in America* (Pope and Tarlov, 1991) and a more recent Institute of Medicine (IOM) disablement model revision highlighted in a report titled *Enabling America: Assessing the Role of Rehabilitation Science and Engineering* (Brandt and Pope, 1997).

The 1991 report uses the original main disablement pathways put forth by Nagi with minor modifications of his original definitions. The

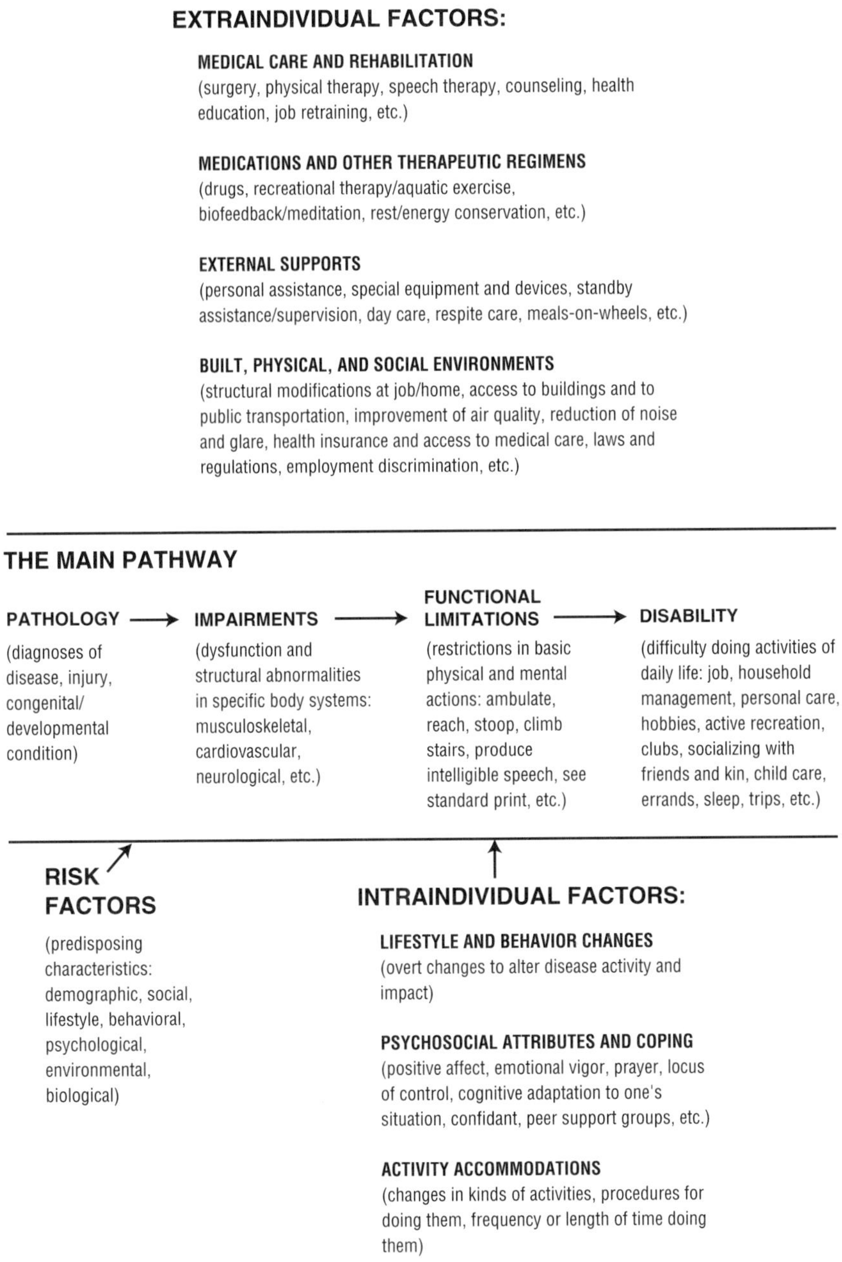

EXTRAINDIVIDUAL FACTORS:

MEDICAL CARE AND REHABILITATION
(surgery, physical therapy, speech therapy, counseling, health education, job retraining, etc.)

MEDICATIONS AND OTHER THERAPEUTIC REGIMENS
(drugs, recreational therapy/aquatic exercise, biofeedback/meditation, rest/energy conservation, etc.)

EXTERNAL SUPPORTS
(personal assistance, special equipment and devices, standby assistance/supervision, day care, respite care, meals-on-wheels, etc.)

BUILT, PHYSICAL, AND SOCIAL ENVIRONMENTS
(structural modifications at job/home, access to buildings and to public transportation, improvement of air quality, reduction of noise and glare, health insurance and access to medical care, laws and regulations, employment discrimination, etc.)

THE MAIN PATHWAY

PATHOLOGY →	IMPAIRMENTS →	FUNCTIONAL LIMITATIONS →	DISABILITY
(diagnoses of disease, injury, congenital/ developmental condition)	(dysfunction and structural abnormalities in specific body systems: musculoskeletal, cardiovascular, neurological, etc.)	(restrictions in basic physical and mental actions: ambulate, reach, stoop, climb stairs, produce intelligible speech, see standard print, etc.)	(difficulty doing activities of daily life: job, household management, personal care, hobbies, active recreation, clubs, socializing with friends and kin, child care, errands, sleep, trips, etc.)

RISK FACTORS
(predisposing characteristics: demographic, social, lifestyle, behavioral, psychological, environmental, biological)

INTRAINDIVIDUAL FACTORS:

LIFESTYLE AND BEHAVIOR CHANGES
(overt changes to alter disease activity and impact)

PSYCHOSOCIAL ATTRIBUTES AND COPING
(positive affect, emotional vigor, prayer, locus of control, cognitive adaptation to one's situation, confidant, peer support groups, etc.)

ACTIVITY ACCOMMODATIONS
(changes in kinds of activities, procedures for doing them, frequency or length of time doing them)

FIGURE 1 The disablement process (Verbrugge and Jette, 1994). Reprinted with permission from Elsevier Science.

1997 report adds two important concepts to the Disablement Model: the concepts of *secondary conditions* and *quality of life*. Both of these concepts are discussed later in this chapter.

In 1997, in an effort to emphasize Nagi's view that disability is not inherent in the individual (as defined by the Social Security Act), but, rather, is a product of the interaction of the individual with the environment, IOM issued *Enabling America*, in which it referred to disablement as "the enabling-disabling process." This effort was an explicit attempt to acknowledge, within the disablement framework itself, that disabling conditions not only develop and progress but can be reversed through the application of rehabilitation and other forms of explicit intervention. Figure 2 is an illustration of Brandt and Pope's 1997 enabling-disabling process.

The Brandt and Pope report (1997) describes the enabling-disabling process as follows:

> Access to the environment, depicted as a square, represents both physical space and social structures (family, community, society). The person's degree of physical access to and social integration into the generalized environment is shown as the degree of overlap of the symbolic person and the environmental square. A person who does not manifest disability (Figure 2a) is fully integrated into society and has full access to both: (1) social opportunities (e.g., employment, education, parenthood, leadership roles) and (2) physical space (e.g., housing, workplaces, transportation). A person with disabling conditions has increased needs (shown as the increased size of the individual) and is dislocated from their prior integration into the environment (Figure 2b). The enabling (or rehabilitative) process attempts to rectify this displacement, either by restoring function in the individual (Figure 2c) or by expanding access to the environment (Figure 2d) (e.g., building ramps). (Brandt and Pope, 1997, p. 3)

International Classification of Impairments, Disabilities, and Handicaps

Independently from the development of the Nagi model, a similar process was also underway in Europe, which led in the early 1970s to the first draft of what later became the World Health Organization (WHO) ICIDH (WHO, 1980). This model also differentiates a series of related concepts: health conditions, impairments, disabilities, and handicaps (WHO, 1980; Badley, 1993). We will refer to these as the ICIDH-1 concepts. ICIDH-1 is not only a conceptual model; it has also associated with it a hierarchical classification of impairment, disability, and handicap (WHO, 1980). We will not review this classification as such, except to note

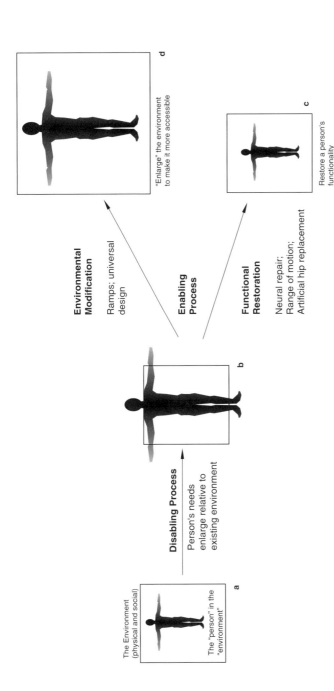

FIGURE 2 Conceptual overview of the enabling-disabling process. The environment, depicted as a square, represents both physical space and social structures (family, community, society). A person who does not manifest a disability (a) is fully integrated into society and "fits within the square." A person with potentially disabling conditions has increased needs (expressed by the size of the individual) and is dislocated from his or her prior integration into the environment (b), that is, "doesn't fit in the square." The enabling (or rehabilitative) process attempts to rectify this displacement, either by restoring function in the individual (c) or by expanding access to the environment (d) (e.g., building ramps) (Brandt and Pope, 1997).

that, in principle, this system provides a scheme for coding and manipulating data on the consequences of health conditions. This classification and the related model of disablement are being revised and have been named ICIDH-2. At the time of this writing (April 1999), a first, beta draft has been circulated for comment (WHO, 1997), and the beta-2 draft is in the final stages of production. The beta-2 draft revised classification will then undergo 2 years of field testing before the final version is prepared for ratification by the WHO. The changes in the definitions and conceptual model that are being recommended in the process of revision to get ICIDH-2 are discussed below. The U.S. National Center for Health Statistics and the Centers for Disease Control and Prevention have served as the lead U.S. agencies in the international ICIDH revision process.

The first component of the ICIDH-1 model is *impairment*, which is defined as follows:

> *In the context of health experience, an impairment is any loss or abnormality of psychological, physiological, or anatomical structure or function.* (WHO, 1980, p. 27)

This definition is similar to Nagi's definition of impairment, but it also includes some of Nagi's notions of pathology. Just as Nagi's impairment is focused on organs or organ systems, impairment as defined here is very much concerned with the function and structure of the body and its components. The ICIDH-2 definition is similar:

> *Impairment is a loss or abnormality of body structure or of a physiological or psychological function.* (WHO, 1997, p. 15)

Huge confusion arises because the ICIDH-1 also uses the word *disability*, but with a slightly different meaning from the Nagi definition of the term. The ICIDH-1 defines *disability* as follows:

> *In the context of health experience, a disability is any restriction or lack (resulting from an impairment) of ability to perform an activity in the manner or within the range considered normal for a human being.* (WHO, 1980, p. 28)

The focus of this definition is very much on the activities carried out by the person. Further understanding of what is included in this definition can be gained by inspection of the associated classification (WHO, 1980, 1997). The activities included range from simple functional activities, such as gripping and holding and maintaining and changing body positions, to more complex activities, such as those related to self-care and other ADLs, IADLs, and some of the activity components of other role activities. The latter include, for example, activities that might be carried out in a work environment. Examples from the ICIDH-1 classification include activities such as organizing a daily routine (ICIDH 1980, Code D18.2), use of foot control mechanisms (ICIDH 1980, Code D67), and tolerance of work stress (ICIDH 1980, Code D76). The ICIDH-1 term *disability* then bridges the Nagi concepts of functional limitation and disability. In revision of the ICIDH, the term *disability* has been replaced by the positive term *activity*, which is defined as follows:

> *Activity is the nature and extent of functioning at the level of the person. Activities may be limited in nature, duration and quality.* (WHO, 1997, p. 14)

To prevent further confusion, the rest of this paper will use the term disability solely in the Nagi sense and use the term *activity limitation* for the ICIDH concept.

In terms of definitions, the construct analogous to the Nagi definition of disability is embodied in the term *handicap*. This is defined as follows:

> *In the context of health experience, a handicap is a disadvantage for a given individual, resulting from an impairment or a disability, that limits or prevents the fulfillment of a role that is normal (depending on age, sex, and social and cultural factors) for that individual.* (WHO, 1980, p. 29)

As is apparent from the definition, *handicap*, like Nagi's *disability*, also embodies the notion of role. However, by referring to disadvantage it goes further than the actual performance of roles to attach a value judgment, that of disadvantage, to restrictions in role performance. The focus of handicap is the person in the society in which he or she lives and reflects cultural norms and expectations for performance.

The term *handicap* did not generally find favor, particularly among people who themselves had disabilities, as it carried within it a history of stigmatization (unrelated to its technical definition). In the ICIDH revision process, this questioning of the term *handicap* spilled over to the whole of the classification and led to the issue of why the emphasis was entirely on the negative. In other words there was a reaction against the whole classification being focused on deficiencies resulting from health conditions. In response to this there has been a switch to neutral terminology, as was illustrated above by the use of the term *activity* instead of the term *disability*. In the proposal for revision of the ICIDH, the concept of handicap, as defined above, has been replaced with the term *participation*, with negative aspects being referred to as *restriction in participation*:

Participation is the nature and extent of a person's involvement in life situations in relation to impairments, activities, health conditions and contextual factors. Participation may be restricted in nature, duration and quality. (WHO, 1997, p. 14)

Like Nagi's definition of disability, the ICIDH definitions of handicap and participation are essentially relational concepts. This is made very explicit in the ICIDH-2, which states that:

> Participation is characterized as the outcome or result of a complex relationship between, on the one hand, a person's health condition, and in particular, the impairments or disabilities he or she may have, and on the other, features of the context that represent the circumstances in which the person lives and conducts his or her life . . . different environments may have a different impact on the same person with impairment or disability. Participation is therefore based on an ecological/environmental interaction model. (WHO, 1997, p. 17)

The conceptual model that accompanies the ICIDH-2 shows that the context potentially has an effect on the expression of all levels of the model: impairment, activity limitation, and restriction in participation. The context refers both to external environmental factors and to more personal characteristics of an individual. The latter range from relatively uncontroversial characteristics, such as age and gender, to aspects of the person relating to educational background, race, experiences, personality and character style, aptitudes, other health conditions, fitness, lifestyle, habits, coping styles, social background, profession, and past and current experience (WHO, 1997). ICIDH-2 includes a draft classification of environmental factors that covers components of the natural environment (weather or terrain), the human-made environment (tools, furnishings, the built environment), social attitudes, customs, rules, practices and institutions, and other individuals (WHO, 1997). All of the above contextual factors may be relevant, in connection with the impairments or activity limitations of a person, for determining whether that person experiences disability in working or not.

Finally, the ICIDH-2 concept of participation goes beyond the performance of roles and deals with the wider issues of the effect of barriers and facilitators to overall participation in society. In the context of work disability these barriers and facilitators include discrimination, stigma, legislation around workplace design and participation (including the Americans with Disabilities Act), attitudes of coworkers, and extra-work issues such as mobility in the community. This means that an assessment of restriction of participation does not necessarily need to be on a personal basis and might, in some situations, be predicted by direct assessment of barriers. For example, workplaces that are not accessible to wheelchair users would systematically restrict participation, irrespective of the nature and demands of the actual work tasks.

CONCEPT OF SOCIAL ROLES

To understand fully how Nagi's definition of disability and the ICIDH definition of handicap can be applied to the area of work disability, one must understand the concept of social role and tasks from a sociological perspective. Social roles, such as being a parent, a construction worker, or a university professor, are basically organized according to how individuals participate in a social system.

According to Parsons (1958), "role is the organized system of participation of an individual in a social system" (p. 316). Tasks are specific activities through which the individual carries out his or her social roles. Social roles are made up of many different tasks, which may be modifiable and interchangeable. For Nagi, the concept of disability is firmly rooted in the context of health. Thus, for Nagi (1991), health-related limitations in the performance of specific social roles are what constitute specific areas of disability, work being one important area of disability. Roles such as work can be disrupted by a variety of factors other than those that are health related. A change in the economic climate or technological changes, for example, may lead to unemployment totally unrelated to health conditions. These would not represent work disability in the way that Nagi defines this term. As Parsons clarifies:

> Roles, looked at that way, constitute the primary focus of the articulation and hence interpretation between personalities and social systems. Tasks on the other hand, are both more differentiated and more highly specified than roles, one role capable of being analyzed into a plurality of different tasks. . . . A task, then, may be regarded as that subsystem of role which is defined by a definite set of physical operations which perform some function or functions in relation to a role. (Parsons, 1958, p. 316)

Are there limits to this concept of disability from the perspective of role performance? Nagi argues that components of roles—expectations or specific tasks that are learned, organized, and purposeful patterns of behavior—are part of the disability concept. They are more than isolated functions or muscle responses (Sarbin and Allen, 1968; Nagi, 1991). Some tasks are role specific, whereas others are common to the enactment of several roles. For Nagi, to the extent that these tasks are learned, organized, and purposeful patterns of behavior, they are part of the disability concept. It is for this reason that Nagi views the concept of disability as ranging from very basic ADLs to the exquisitely complex social roles such as one's occupation. Since activities of daily living (e.g., dressing, bathing, and eating) are part of a set of expectations inherent in a variety of other social roles, Nagi sees deviations or limitations in the performance of even such basic social roles as components of the concept of disability (Nagi, 1991). For Nagi, disability as a heuristic concept is inclusive of all socially defined roles and tasks.

In the ICIDH-2, overall role performance mainly falls into the domain of participation. The boundary between activity limitation and participation is drawn differently from the way in which it is drawn in the Nagi model, in that a person who is unable to perform activities that are the components of roles is considered to have activity limitations (Figure 3). These are the roles that Nagi refers to as "basic social roles." In the context of work disability, the distinction is between restriction of participation related to work as an overall concept and the carrying out of the activities involved in the work itself. This is discussed in more detail in the section that explores conceptual issues related to work disability.

Fundamental to differentiating the concept of disability from those of pathology, impairment, and functional limitation is the consideration of the difference between concepts of attributes or properties on the one hand and relational concepts on the other (Cohen, 1957).

As Nagi describes it:

> Concepts of attributes and properties refer to the individual characteristics of an object or person, such as height, weight, or intelligence. Indicators of these concepts can all be found within the characteristics of the individual. Pathology, impairment, and functional limitations are concepts of attributes or properties. . . . Disability is a relational concept; its indicators include individuals' capacities and limitations, in relation to role and task expectations, and the environmental conditions within which they are to be performed. (Nagi, 1991, p. 317)

Let us take the example of limitation in the performance of one's work role—or work disability. Work disability typically begins with the onset of one or more health conditions that *may* limit the individual's performance of specific tasks through which an individual would typically perform his or her job. The onset of a specific health condition—say, a stroke or a back injury—may or may not lead to actual limitation in performing the work role, a work disability. The development of work disability will depend, in part, on the extent to which the health condition limits the individual's ability to perform specific tasks that are part of one's occupation, and alternatively, degree of work disability may depend on external factors, for example, workplace attitudes—say, flexible working hours—that may restrict employment opportunities for persons with specific health-related limitations. Or work disability might be affected by

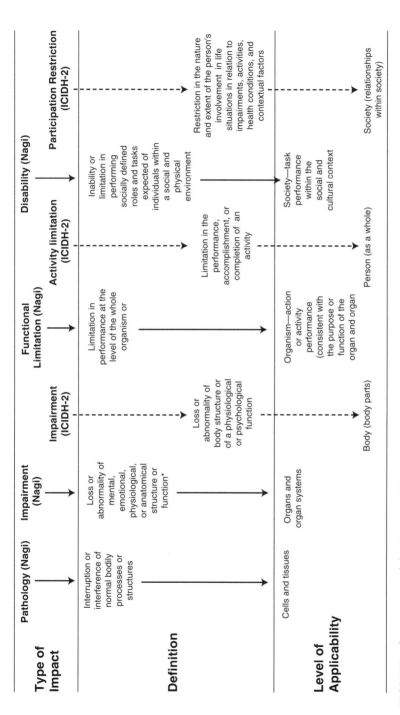

FIGURE 3 Comparison of the Nagi and ICIDH-2 concepts and definitions.
*Includes all losses and abnormalities, not just those attributable to active pathology; also includes pain.

accessible modes of transportation to the workplace, environmental barriers in the workplace, or the willingness of the employer to modify the individual workstation to accommodate a health condition. Viewed from the perspective of role performance, degree of work disability could be reduced by improving the individual's capacity to accomplish functional activities—a very traditional view of rehabilitation—or by manipulating the physical or social environment in which work occurs. A discussion similar to that given above could be formulated by using the language of the ICIDH.

The fundamental conceptual issue of concern is that a health-related restriction in work participation may not be solely or even primarily related to the health condition itself or its severity. In other words, although the presence of a health condition is a prerequisite, "work disability" may be caused by factors external to the health condition's impact on the structure and functioning of a person's body or the person's accomplishment of a range of activities.

DIRECTIONALITY AND THE DYNAMIC NATURE OF DISABILITY

The earliest disablement models represented by the ICIDH-1 formulation (WHO, 1980) and Nagi's disablement model (Nagi, 1965) presented the disablement process as more or less a simple linear progression of response to illness or consequence of disease. One consequence of this traditional view is that disabling conditions have been viewed as static entities (Marge, 1988). This traditional, early view of disablement failed to recognize that disablement is more often a dynamic process that can fluctuate in breadth and severity across the life course. It is anything but static or unidirectional.

More recent disablement formulations or elaborations of earlier models have explicitly acknowledged that the disablement process is far more complex (Pope and Tarlov, 1991; Verbrugge and Jette, 1994; Brandt and Pope, 1997; WHO, 1997; Fougeyrollas, 1998). These more recent authors all note that a given disablement process may lead to further downward-spiraling consequences. These feedback consequences, which may involve pathology, impairments, and further limitations in function or disability, have been explicitly incorporated into the graphic illustrations of more recent disablement formulations. The Pope and Tarlov (1991) report uses the term *secondary conditions* to describe any type of secondary consequence of a primary disabling condition. Commonly reported secondary conditions include pressure sores, contractures, depression, and urinary tract infections (Marge, 1988); but it should be understood that they can be either a pathology, an impairment, a functional limitation, or an additional disability.

Longitudinal analytic techniques now exist to incorporate secondary conditions into research models and are beginning to be used in disablement epidemiological investigations (Lawrence and Jette, 1996).

HOW DISABLEMENT CONCEPTS DIFFER FROM QUALITY OF LIFE AND SIMILAR CONCEPTS

To compare disablement concepts with the phenomenon of quality of life, one must first consider how quality of life has been defined in the literature. Birren and Dieckermann have provided a useful starting point:

> The concept of quality of life is complex, and it embraces many characteristics of the social and physical environments as well as the health and internal states of individuals. There are two approaches to the measurement of quality of life: One is based upon the subjective or internal self perceptions of the quality of life; the other approach is objective and based upon external judgments of the quality of life. (Birren and Dieckermann, 1991, p. 350)

If we apply Birren and Dieckermann's perspective to work roles and work disability, objective dimensions of quality of life might include whether a person has had to change jobs because of a health problem, whereas the subjective dimension might include the individual's satisfaction with his or her job. Consistent with this objective and subjective view of quality of life, Lawton (1983) has suggested that measures of quality of life should include a multidimensional evaluation of both intrapersonal and social-normative criteria including:

1. psychological well-being,
2. perceived quality of life,
3. behavioral competence in multiple areas (i.e., health, functional health, cognition, time use, and social behavior), and
4. the objective environment itself.

Indicators of quality of life are extremely broad and have included standard of living, economic status, life satisfaction, quality of housing and the neighborhood in which one lives, self-esteem, and job satisfaction. Such a broad concept subsumes many dimensions of personal well-being not directly related to health.

In response to concerns about the breadth of overall quality of life, some health researchers have adopted a narrower concept called "health-related quality of life." Health-related quality of life has been defined in line with WHO's definition of health as a state of complete physical, mental, and social well-being, not merely the absence of disease or infirmity (WHO, 1947). Major dimensions in the health-related quality-of-life measures include signs and symptoms of disease, performance of basic physical activities of daily life, performance of social roles, emotional state, intellectual functioning, general satisfaction, and perceived well-being.

Some models of disablement such as the IOM formulation (Pope and Tarlov, 1991; Brandt and Pope, 1997) and Patrick's (1997) conceptual work clearly define quality of life as distinct from the disabling process. As Pope and Tarlov (1991) describe it:

> Quality of life affects and is affected by the outcomes of each stage of the disabling process. Within the disabling process, each stage interacts with an individual's quality of life; it is not an endpoint of the model but rather an integral part. (p. 8)

This view of quality of life strikes the authors as inconsistent with the definitions of quality of life described previously and may create problems in designing appropriate survey measures. The concepts of quality of life and health-related quality of life, in particular, appear to overlap and include within their boundaries many (yet certainly not all) of the disablement concepts reviewed in this paper. Like the disablement concept, quality of life includes dimensions at the personal activity and social role levels. Like the disablement concepts, quality of life does direct some attention to the concepts of disease, through an assessment of signs and symptoms. Most quality-of-life measures focus little attention on organ and body system functioning and focus more on the consequences of impairments at the personal activity or social role level. At the level of social roles, quality-of-life dimensions are broader than the disablement concepts that incorporate overall life satisfaction, energy, vitality, and emotional well-being (Levine and Croog, 1984).

Thus, the authors have difficulty viewing the concept of quality of life as entirely distinct from several dimensions in the disablement concepts. For some elements of quality of life, disablement is clearly a precursor, but other elements fall outside the disablement formulation. There appears to be considerable overlap between elements of the two formulations, and a conceptualization that acknowledges this overlap may be a more useful formulation (Figure 4).

CONCEPTUAL ISSUES RELATED TO THE MEASUREMENT OF WORK DISABILITY

The underlying structure of models of disablement, as currently conceived, maps a pathway between the health condition and the ensuing

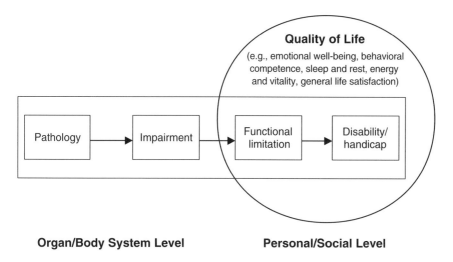

FIGURE 4 Relationship of disablement process to quality of life.

"work disability" or other restrictions to social participation. Close inspection of the definitions given above suggests that a number of steps can be identified in the pathway between the health condition and the social consequences described as work disability. At a micro level there are the pathological changes in the body and impairment in the structure and functioning of organs and body systems. There may be an impact on the activity of the person, ranging from simple movements, to basic activities of daily living, to instrumental activities of daily living, and so on. These can then contribute to the performance of more complex social roles, and ultimately, the person's participation in all aspects of society can be adversely affected. Work is one such social role.

Both the Nagi and the ICIDH models cover the spectrum of the consequences of health conditions. As indicated earlier, as well as terminology, a major difference is where these models place the boundaries between the different concepts (see Figure 3). In the Nagi model the performance of all activities, except for basic actions or functions of the body, is subsumed into the overall category of disability (Nagi, 1976). In the ICIDH model the concept of activity includes these basic actions as well as ADLs, IADLs, and some other role activities (with the emphasis very much on activity) (WHO, 1980, 1997; Badley, 1993). Participation is reserved to highlight the way in which the performance of activities may be constrained by more than the immediate context of the activity. The juxtaposition of the two models in this way illuminates some nuances in the ways in which the impact of health conditions can been conceptualized as having an impact on the overall functioning of the individual. As indicated earlier, work disability is a function of whether the person can perform specific work-related tasks and of external factors. From the point of view of the measurement of work disability, it may be useful to distinguish between the degree of difficulty that a person may have in carrying out an activity and these other factors (such as barriers in the environment, attitudes of employers or coworkers, and other restrictions) that might prevent the performance of those activities in daily life. In this way, the levels of impact described within the conceptual models are of importance as they allow us to locate where many of the current types of assessment of work disability might fit in.

In the authors' view, in general, no explicit conceptual framework appears to be used in the ascertainment of work disability. A number of implicit conceptual approaches appear to have been used to assess and identify people with possible work disabilities. Each approach can be compared to the different levels of a model of disablement as discussed in the previous sections. We will review these in turn. However, before we do this we need to deal with some more general issues.

Discrete or Continuous Phenomena

Disability is commonly presented as an all-or-nothing phenomenon; either a person has a disability or a person does not. In reality, disability (in particular, roles or activities) is usually encountered in terms of degree of difficulty, limitation, or dependence, ranging from slight to severe. The question then becomes: where on the disability spectrum is that threshold that determines if a person is disabled? This needs to take into account any assistive devices or accommodations that the person may have. In the current context, work participation is often determined as being an endpoint, in that either people have a work disability or they do not. In reality, the situation is likely to be more complex. For example, many people with functional and activity limitations may continue to work, but their labor force participation may be compromised in some way by the condition. To the extent that it is, these people might be said to have some degree of work disability. In measuring work disability, a clear definition of the threshold used needs to be made. Alternatively, a continuous measurement needs to be undertaken.

Duration or Chronicity

There is a pervasive assumption that work disability is long-term state. Stereotypes about disability are dominated by the archetype of a person who uses a wheelchair. Embedded in this is the notion of some disabling event, a period of adjustment and rehabilitation, and then the resumption of as full a life as possible with the assistance of any necessary assistive devices or accommodations. With many impairments, the reality of disability is somewhat different. The majority of individuals in the working-age population with long-term activity restrictions report that this restriction is due to musculoskeletal, circulatory, or respiratory disorders (LaPlante and Carlson, 1996). These conditions may also be associated with varying degrees of "illness," so that it is not just an issue of physical performance. There are also considerations of pain, fatigue, and other symptoms. Many of these conditions are episodic in nature and may have trajectories of either deterioration or recovery (the latter being less common). This means that, apart from any environmental barriers or facilitators, the day-to-day or month-to-month experience of disability may be variable. This may need to be taken into account in any measurement scheme.

Examples of Conceptual Approaches to Measuring Work Disability

All disablement concepts appear to have been addressed, at least to some extent, as part of efforts to assess work disability.

Health Condition or Pathology

Under some circumstances knowledge of the health condition or pathology contributes to an assessment of work disability. Medical listings of diagnosis and medical severity have been used by some agencies to identify individuals who would be unlikely to benefit from vocational rehabilitation (Reno, 1999). Such listings have also been applied in the context of Social Security disability determinations. Concern has been expressed because the use of such listings might, on the one hand, deny benefits to individuals who need them and might, on the other, award benefits to those who could still work. Such concerns are a reflection of the many steps in the disablement model between the health condition and work disability.

Impairment

Assessments of work disability, or at least of entitlement to compensation for work injury, are often made at the level of impairment. The classic assessment is perhaps what has been pejoratively referred to as the "meat chart" assessment of the consequences of traumatic injury. An example of this would be the American Medical Association *Guides to the*

Evaluation of Permanent Impairment (American Medical Association, 1993), which is a standardized system for translating the extent of an injury of a body part into a percentage of disability of the whole person. This type of system has been used for the assessment of compensation payments, including for workers' compensation.

A number of assessments focus on the functioning of the body, for example, assessments of strength, muscular endurance, body coordination and flexibility, and cognitive and sensory functions (Fleishman, 1972, 1999). The problem with this impairment-focused approach is that even though these assessments may be made in the context of relating functional requirements with the requirements of certain jobs, one needs empirical evidence to support the contention that the degree of impairment is going to have a direct relationship to work disability. Without such evidence, the validity of such an approach is highly suspect.

Functional Limitation

Much of the discussion of assessment of work effectively has been at the level of functional disability. An example would be the assessment of abilities proposed for the Occupational Information Network (O*NET) system (see, for example, Wunderlich, 1999, p. 24). Here abilities such as oral comprehension, memorization, finger dexterity, and depth perception (Wunderlich, 1999, p. 35) will be assessed and compared with the average requirements of particular jobs. Although the intent was that this should be done for all jobs, it has been suggested that this approach could, in principle, provide the basis of an assessment of work disability (Wunderlich, 1999, p. 86). Measures of work-related functional capacity (Lechner et al., 1997) have also been devised to test or ask about activities such as lifting, standing, walking, sitting, and carrying. Although closer in concept to work disability than assessments of pathology and impairment, assessments of capacity to perform work functions are one level removed from the concept of work disability. They look at the specific abilities of the individual for work in standardized ways not directly related to actual work settings. More importantly, they take no account of any environmental barriers or facilitators that might moderate the way in which a person's functional limitations are expressed as disabilities.

Activity Limitation (at Work)

A direct way of answering at least part of the question about work disability is to carry out a workplace assessment. This gives information about whether the person can actually carry out the requirements for the major components of the job. This is the kind of assessment that is frequently carried out in the context of vocational rehabilitation. However, factors other than the actual performance of the work tasks likely contribute to work disability as indicated earlier. This is further discussed below.

Work Disability

Having separated out the activity limitation in work tasks, one can look at work disability from the perspective of carrying out a work role. Direct assessment of work disability involves several elements related to the role of work. These include:

- activities within the workplace;
- a range of other aspects including necessary mobility in getting to work;
- interaction with colleagues, superiors, and subordinates; and
- the amount and type of work that can be carried out.

Work disability is most frequently assessed by direct inquiry of the individual. The measurement problems with this kind of approach are reviewed in Chapter 3. In population surveys the two main types of approaches to measurement of work disability are either (1) direct questioning about any limitations in work attributable to a health condition or (2) the independent ascertainment of disability and work status, with some inference of a connection between disability and work status. We will review each of these in turn.

DIRECT ASSESSMENT OF WORK DISABILITY

The most direct approach to ascertainment of work disability is to inquire about working status together with questions as to whether non-participation is health related. There are various permutations on these types of questions. Some typical formulations are shown in Figure 5.

As Figure 5 illustrates, typical survey questions about work disability are asked with a general reference to work, and it is left to the respondent to determine the specific relevant elements to be considered within the work role. If the respondent is currently working or has recently worked, this is presumably taken to mean the most recent working experience. If the person is not working, then this is more problematic. The answer to the question will depend on what type of employment, if any, the individual has in mind when answering the question. If the purpose of the question is to determine incapacity for work, then the nature of the job and any accommodations that have been or might be made is crucial. Few

1990 Decennial Census: Work Disability
Does this person have a physical, mental or other health condition that lasted for 6 months or more which
(a) limits the kind or amount of work this person can do at a job?
(b) prevents this person from working at a job?

U.S. Census for Year 2000
General question about activity limitations (difficulty in carrying out specific activities) because of a physical, mental, or emotional condition lasting 6 months or more.

March Current Population Surveys, 1981–1988
The CPS has a set of criteria. If one or more of the final four conditions was met, the person was considered to have a severe work disability:
1. Does anyone in the household have a health problem or disability which prevents them from working or which limits the kind or amount of work they can do?
    Is there anyone in this household:
2. Who ever retired or left a job for health reasons?
3. Did not work in the survey week because of a long-term physical or mental illness or disability which prevents the performance of any kind of work?
4. Did not work at all in the previous year because ill or disabled?
5. Under 65 years of age and covered by Medicare?
6. Under 65 years of age and a recipient of Supplemental Security Income (SSI)?

Survey of Income and Program Participation (Third Wave Supplement), 1984
Does _____'s health or condition limit the kind or amount of work _____ can do?

National Health Interview Surveys
Phase 1
a. Does _____'s health now keep him from working?
b. Is he limited in the kind of work he could do because of his health?
c. Is he limited in the amount of work he could do because of his health?
d. Is he limited in the kind or amount of other activities because of his health?
Phase 2
a. Does ____ now have a job?
b. In terms of health is ____ now able to work?
c. Is he limited in the kind of work he could do because of his health?
d. Is he limited in the amount of work he could do because of his health?
e. Is he limited in the kind or amount of other activities because of his health?

FIGURE 5 Examples of survey questions.

survey research approaches break down work role into its major component parts to determine the perceived degree of disability within each.

Typical survey research questions also leave it to the respondent to
attribute not working to an underlying health condition. It may be that
the individual answers that he or she cannot work, yet the person may not
be given the opportunity to specify the circumstances under which this
might be possible. A survey of working-age people with disability in the
United States showed that over two-thirds wanted to work (Stoddard et
al., 1988, p. 24). In the 1991 Canadian Health and Activity Limitation
Survey, 64 percent of respondents with disabilities reported that they
were not in the labor force, and over two-thirds of these said that they
were completely prevented from working (Statistics Canada, 1993). However, all respondents were given the opportunity to answer questions
about needed accommodations in the workplace. Despite reporting that
they were prevented from working, 69 percent of these individuals
reported needing a variety of workplace accommodations (such as job
redesign or modified hours) and 76 percent reported needing adaptations
(such has handrails, elevators, or modified workstations). Whether or not
the provision of such accommodations or adaptations would facilitate
workplace reintegration is unknown. However, the findings illustrate how
changing the framing of a question sheds a different light on what it
means to be unable to work. Individuals who were not in the labor force
were also asked about barriers to employment. The most frequently mentioned barriers were losing some or all of their current income, feelings
that their training was not adequate, no available jobs, and loss of additional supports (such as health benefits). Other less frequently mentioned
reasons were family responsibilities, having being the victim of discrimination, and not having accessible transportation (Statistics Canada, 1993).
In other words, most of the reasons were related not to the nature of the
work, but to some of the other circumstances surrounding the issue of
work disability.

Furthermore, some individuals will have a choice as to how they
describe their working status. For example, a person with a disability who
also has small children could variously describe him- or herself as a home-
maker or not being in the labor force because of the disability. Or people
leaving the workforce in their 50s may describe themselves as having
taken an early retirement. Without extra information it may be difficult to
tell whether this is indeed the situation or whether the alternative
description was seen as a less stigmatizing alternative to describing themselves as having a work disability.

In a survey research situation, if a person is working, the typical
approach is to assume that no work disability is present. Nevertheless,
the person may be limited in the amount or kind of work done or both.

The person may be spending less time working, working at a less skilled job, or earning less money. This information can be obtained from survey questions (see Figure 5), but often with relatively little qualifications as to what this means. What is less often addressed is that for many people with disabilities working may mean forgoing opportunities to participate in other areas of life. Just going to work may, for example, exhaust all reserves of energy or require time-consuming preparations. There is a fine line between what might be considered a satisfactory accommodation and an unsatisfactory compromise or necessity, and different people will value this trade-off differently.

CONCLUSION

The problem with all the approaches to work disability, as indicated by our discussion of conceptual frameworks, is that there is unlikely to be a one-to-one relationship between the presence of health conditions, impairments, functional limitations, or activity restrictions and disability in employment. There is a pervasive assumption that work disability relates to the person's degree of functional limitation and activity restriction. This is reflected in the concern about assessment, where the focus is very much on the individual's performance. Lip service is paid to the environment, particularly in the context of work disability and vocational rehabilitation. As we have tried to show, a full understanding of work disability needs to take into account the individual's circumstances and the social and physical environments of the workplace.

The research challenge is to apply the insights provided by the models of disablement to come to a common understanding of work disability and to understand the relationships and the dynamics of the pathway between health conditions and work disability. Researchers need to find ways to incorporate an understanding of the external factors that influence the development of work disability into its measurements.

REFERENCES

American Medical Association. 1993. *Guides to the Evaluation of Permanent Impairment.* 4th ed. Chicago: American Medical Association.
Badley EM. 1993. An introduction to the concepts and classifications of the International Classification of Impairments, Disabilities, and Handicaps. *Disability and Rehabilitation* 15:161–178.
Badley EM. 1995. The genesis of handicap: definitions, models of disablement, and role of external factors. *Disability and Rehabilitation* 17:53–62.
Birren J, Dieckermann L. 1991. Concepts and content of quality of life in the later years: an overview. In: Birren J, et al., eds. *Quality of Life in the Frail Elderly.* New York, NY, Academic Press, Inc. Pp. 344–360.

Brandt E, Pope A, eds. 1997. *Enabling America: Assessing the Role of Rehabilitation Science and Engineering.* Washington, DC: National Academy Press.

Cohen MA. 1957. *Preface to Logic.* New York: Meridian Books.

Fleishman EA. 1972. Structure and measurement of psychomotor abilities. In: Singer RN, ed. *The Psychomotor Domain: Movement Behavior.* Philadelphia: Lea and Febiger.

Fleishman EA. 1999. Linking components of functional capacity domains with work requirements. In: Wunderlich GS, ed. *Measuring Functional Capacity and Work Requirements: Summary of a Workshop.* Washington DC: National Academy Press.

Fougeyrollas P, ed. 1998. *ICIDH and Environmental Factors International Network.*Volume 9, Numbers 2–3. Quebec, Canada: Canadian Society for the ICIDH.

Jette A, Assmann S, Rooks D, Harris B, Crawford S. 1998. Interrelationships among disablement concepts. *J of Gerontol: Med Sci* 53A(5):M395–M404.

LaPlante MP, Carlson D. 1996. *Disability in the United States: Prevalence and Causes, 1992. Disability Statistics. Report 7.* Washington, DC: U.S. Department of Education, National Institutes on Disability and Rehabilitation Research.

Lawrence R, Jette A. 1996. Disentangling the disablement process. *Journal of Gerontol Soc. Sciences* 51B(4):S173–S182.

Lawton MP. 1983. Environment and other determinants of well being in older people. *Gerontologist* 23:349–357.

Lechner, D, Roth D, Straaton K. 1997. Functional capacity evaluation in work disability. *Work* 1:31–47.

Levine S, Croog S. 1984. What constitutes quality of life? A conceptualization of the dimensions of life quality in healthy populations and patients with cardiovascular disease. In: Wenger N, et al., eds. *Assessment of Quality of Life in Clinical Trials of Cardiovascular Therapies.* New York, NY: LeJacq Publ. Co. Pp. 46–66.

Marge M. 1988. Health promotion for people with disabilities: Moving beyond rehabilitation. *AM J Health Promotion* 2(4):29–44.

Mathiowetz N. 2000. Methodological issues in the measurement of work disability. In: Mathiowetz N, Wunderlich, GS, eds. *Survey Measurement of Work Disability: Summary of a Workshop.* Washington, DC: National Academy Press. Pp. 28–52.

Nagi S. 1964. A study in the evaluation of disability and rehabilitation potential: Concepts, methods, and procedures. *American Journal of Public Health* 54:1568–1579.

Nagi S. 1965. Some conceptual issues in disability and rehabilitation. In: *Sociology and Rehabilitation.* Sussman MB, ed. Washington, DC: American Sociological Association.

Nagi S. 1976. An epidemiology of disability among adults in the United States. In: *Health & Society. Milbank Memorial Fund Quarterly* 54:439–467.

Nagi S. 1991. Disability concepts revisited: Implications for prevention. In: Pope A, Tarlov A, eds. *Disability in America: Toward a National Agenda for Prevention.* Washington, DC: National Academy Press.

Parsons T. 1958. Definitions of health and illness in the light of American values and social structure. In: Jaco EG, ed. *Patients, Physicians, and Illness.* Glencoe, Ill: Free Press.

Patrick DL. 1997. Rethinking prevention for people with disabilities: A conceptual model for promoting health. *American Journal of Health Promotion* 11(4):257–260.

Pope A, Tarlov A, eds. 1991. *Disability in America: Toward a National Agenda for Prevention.* Washington, DC: National Academy Press.

Reno V. 1999. Adapting measurement of functional capacity to work to SSA's disability decision process. In Wunderlich GS, ed. *Measuring Functional Capacity and Work Requirements. Summary of a Workshop.* Washington, DC: National Academy Press. Pp. 74–78

Sarbin T, Allen V. 1968. Role theory. In: Linsey G, Aronson E, eds. *The Handbook of Social Psychology,* Vol. 1, 2nd edition. Reading MA: Addison-Wesley Publ. Co.

Statistics Canada. 1993. Adults with disabilities: Their employment and education charac-
    teristics. 1991 Health and Activity Limitation Survey. Ottawa: *Statistics Canada Cata-
    logue 82–554.*
Stoddard S, Jans L, Ripple J, Kraus L. 1998. *Chartbook on Work and Disability in the United
    States, 1998. An InfoUse Report.* Washington, DC: U.S. National Institute on Disability
    and Rehabilitation Research.
Verbrugge L, Jette A. 1994. The disablement process. *Soc Sci & Med.* 38(1):1–14.
World Health Organization (WHO). 1947. *Constitution of the World Health Organization.* New
    York, NY: WHO.
WHO. 1980. *The International Classification of Impairments, Disabilities, and Handicaps—A
    Manual Relating to the Consequences of Disease.* Geneva: WHO.
WHO. 1997. *ICIDH-2: International Classification of Impairments, Activities, and Participation. A
    Manual of Dimensions of Disablement and Functioning. Beta-1 Draft for Field Trials.* Geneva:
    WHO.
Wunderlich GS, ed. 1999. *Measuring Functional Capacity and Work Requirements.* Washington,
    DC: National Academy Press.

Methodological Issues in the Measurement of Work Disability[1]

Nancy A. Mathiowetz, Ph.D.[2]

The collection of information about persons with disabilities presents a particularly complex measurement issue because of the variety of conceptual paradigms that exist, the complexity of the various paradigms, and the numerous means by which alternative paradigms have been operationalized in different survey instruments (see paper by Jette and Badley for a review). For example, disability is often defined in terms of environmental accommodation of an impairment; hence, two individuals with the same impairment may not be similarly disabled or share the same perception of their impairment. For an individual with mobility limitations who lives in an assisted-living environment that accommodates the impairment, the environmental adaptations may result in little or no disability. The same individual living on the second floor of an apartment building with no elevator may have a very different perception of the impairment and may see him- or herself as disabled because of the environmental barriers that exist within his or her immediate environment.

The Social Security Administration (SSA) is currently reengineering its disability claims process for providing benefits to blind and disabled

---

[1]This paper was originally prepared for the committee workshop titled "Workshop on Survey Measurement of Work Disability: Challenges for Survey Design and Method" held on May 27–28, 1999 in Washington, D.C. (IOM, 2000).

[2]Nancy Mathiowetz is an Associate Professor at the University of Maryland's Joint Program in Survey Methodology.

persons under the Social Security Disability Insurance (SSDI) and Supplemental Security Income (SSI) programs. As part of the effort to redesign the claims process, SSA has initiated a research effort designed to address the growth in disability programs, including the design and conduct of the Disability Evaluation Study (DES). The DES will provide SSA with comprehensive information concerning the number and characteristics of persons with impairments severe enough to meet SSA's statutory definition of disability, as well as the number and characteristics of people who are not currently eligible but who could be eligible as a result of changes in the disability decision process. For those years in which the DES is not conducted, SSA will need to monitor the potential pool of applicants. One means by which SSA can monitor the size and characteristics of potential beneficiaries is through other ongoing federal data collection efforts. For both the conduct of the DES and the monitoring of the pool of potential beneficiaries through the use of various data collection efforts, it is critical to understand the measurement error properties associated with the identification of persons with disabilities as a function of the essential survey conditions under which the data have been and will be collected. The extent to which alternative instruments designed to measure persons with disabilities map to various eligibility criteria under consideration by SSA is also important.

BACKGROUND

The collection of disability data is an evolving field. Although a large and growing number of scales attempt to measure functional status and work disability, little is known about the measurement error properties of various questions and composite scales. The empirical literature provides clear evidence of variation in the estimates of the number of persons with disabilities in the United States, depending upon the conceptual paradigm of interest, the analytic objectives of the particular measurement process, and the essential survey conditions under which the information is collected (e.g., Haber, 1990; McNeil, 1993; Sampson, 1997). This literature suggests that estimates of the disabled population not only are related to the conceptual framework underlying the measurement construct but are also a function of the essential survey conditions under which the measurement occurred, including the specific questions used to measure disability, the context of the questions, the source of the information (self- versus proxy response), variations in the mode and method of data collection, and the sponsor of the data collection effort. Furthermore, terms such as *impairment, disability, functional limitation,* and *participation* are often inconsistently used, resulting in different and conflicting estimates of prevalence. Attempts to measure not only the prevalence but also the severity of an impairment or disability further complicate the measurement process.

Recent shifts in the conceptual paradigm of disability, in which disability is viewed as a dynamic process rather than a static measure and as an interaction between an individual with an impairment and the environment rather than as a characteristic only of the individual, imply that those responsible for the development of disability measures must separate the measurement of the impact of environmental factors in the enablement-disablement process from the measurement of ability. Viewing disability as a dynamic state resulting from an interaction between a person's impairment and a particular environmental context further complicates the assessment of the quality of various survey measures of disability, specifically, the reliability of a measure. As a dynamic characteristic, one would anticipate changes in the reports of disability as a function of changes in the individual as well as changes in the social and environmental contexts. The challenge for the measurement process is to disentangle true change from unreliability.

This workshop comes at a time when the federal government is undertaking several initiatives with respect to the measurement of disability in federal data collection efforts. The Americans with Disability Act of 1990 (ADA) defines disability as (1) a physical or mental impairment that substantially limits one or more of the major life activities of the individual, (2) a record of a substantially limiting impairment, or (3) being regarded as having a substantially limiting impairment. Although the measurement of disability within household surveys is not bound by the ADA definition, the passage of the ADA provides a socioenvironmental framework for how society comprehends and uses terms such as *disability* and *impairment* (e.g., the popular press and court rulings on ADA-related litigation). These definitions will evolve as a function of litigation related to ADA legislation and presentation of that litigation in the press. Hence, society is entering a period in which potential dynamic shifts in the comprehension and interpretation of the language associated with the measurement of persons with disabilities can be anticipated.

This paper is intended to serve as a means of facilitating discussion among individuals from diverse theoretical and empirical disciplines concerning the methodological issues related to the measurement of persons with disabilities. As a first step to achieving this goal, a common language and framework needs to be established for the enumeration and assessment of the various sources of error that affect the survey measurement process. The chapter draws from several empirical investigations to provide evidence as to the extent of knowledge concerning the error properties associated with various approaches to the measurement of functional limitations and work disability.

SOURCES OF ERROR IN THE SURVEY PROCESS: THE SURVEY RESEARCH PERSPECTIVE

For the purpose of defining a framework that can be used to examine error associated with the measurement of persons with disabilities, I draw upon the conceptual structure and language used by Groves (1989), based on earlier work of Kish (1965) and used by Andersen et al. (1979). Suchman and Jordan (1990) have described errors in surveys as the discrepancy between the concept of interest to the researcher and the quantity actually measured in the survey. Bias, according to Kish (1965, p. 509), refers to systematic errors in a statistic that affect any sample taken under a specified survey design with the same constant error or, as stated by Groves (1989), is the type of error that affects the statistic in all implementations of a survey. Variable errors are those errors that are *specific* to a particular implementation of a design, that is, specific to the particular trial. The concept of variable error requires the possibility of repeating the survey, with changes in the units of replication, that is, the particular set of respondents, interviewers, supervisors, coding, editing, and data entry staff.

Errors of Nonobservation

Within the framework of survey methodology, both variable error and bias are further characterized in terms of errors of nonobservation and errors of observation. As one would expect from the term, errors of nonobservation reflect failure to obtain observations for some segment of the population or for all elements to be measured. Errors of nonobservation are most often classified as arising from three sources: sampling, coverage, and nonresponse.

Sampling Error

Sampling error represents one type of nonobservation variable error; it arises from the fact that measurements (observations) are taken for only a subset of the population. Sampling variance refers to changes in the value of some statistic over possible replications of a survey in which the sample design is fixed but different individuals are selected for the sample. Estimates based on a particular sample will not be identical to estimates based on a different subset of the population (selected in the same manner) or to estimates based on the full population.

Coverage Error

Coverage error defines the failure to include all eligible population members on the list or frame used to identify the population of interest. Those members not identified on the frame have a zero probability of selection and are never measured. For example, in the United States, approximately 5 percent of the population lives in households without telephone service; any survey that is conducted by telephone and that attempts to describe the entire household-based population of the United States therefore suffers from coverage error. To the extent that those without telephones differ from those with telephones for the construct of interest, the resulting estimates will be biased.

Nonresponse Error

Nonresponse error can arise from failure to obtain any information from the persons selected to be measured (unit nonresponse) or from failure to obtain complete information from all respondents to a particular question (item nonresponse). The extent to which nonresponse affects survey statistics is a function of both the rate of nonresponse and the difference between respondents and nonrespondents, as illustrated in the following formula:

$$y_r = y_n + \left(\frac{nr}{n}\right)(y_r - y_{nr})$$

where:

y_r = the statistic estimated from the r respondents,
y_n = the statistic estimated from all n sample cases,
y_{nr} = the statistic estimated from the nr nonrespondents, and
nr = the proportion of nonrespondents.

Knowing the response rate is not sufficient to determine the level of nonresponse bias; studies with both high and low rates of nonresponse can suffer from nonresponse bias.

As noted by Groves and Couper (1998), it is useful to further distinguish among the types of unit nonresponse, each of which may be related to the failure to measure different types of persons. For most household data collection efforts involving interviewers, the final outcome of an interview attempt is often classified into one of the following four categories: completed or partial interview, refusal, noncontact, and other noninterview.[3] Survey design features can affect the distribution of cases across the various categories. Noncontact rates are affected by the length of the field period (in which short field periods result in higher noncontact rates than longer field periods). Surveys that place greater demands on the respondent may suffer from higher refusal rates than less burdensome instruments. The choice of respondent rule affects the rate of nonresponse; designs that permit any knowledgeable adult within the household to serve as the respondent provide an interviewer with some flexibility, should one adult within the household refuse or be unable to participate. Field efforts that fail to accommodate non-English-speaking respondents or that focus their attention on frail subpopulations tend to experience higher rates of other noninterviews.

Errors of Observation

Observational errors can arise from any of the elements directly engaged in the measurement process, including the questionnaire, the respondent, and the interviewer, as well as the characteristics that define the measurement process (e.g., the mode and method of data collection). This section briefly reviews the theoretical framework and empirical findings related to the various sources of measurement error in surveys.

Questionnaire as Source of Measurement Error

Tourangeau (1984) and others (see Sudman et al. [1996] for a review) have categorized the survey question-and-answer process as a four-step process involving comprehension of the question, retrieval of information from memory, assessment of the correspondence between the retrieved information and the requested information, and communication of the response. In addition, the encoding of information, a process outside the control of the survey interview, determines a priori whether the information of interest is available for the respondent to retrieve.

Comprehension of the question involves the assignment of meaning to the question by the respondent. Ideally, the question will convey the meaning of interest to the researcher. However, several linguistic, structural, and environmental factors affect the interpretation of the question by the respondent. These factors include the specific wording of the ques-

---

[3]*Other noninterview* is used to classify cases in which contact was made with the members of the household in which the sample person resides, but for reasons such as physical or mental health, language difficulties, or other reasons not related to reluctance to participate, the interviewer was unable to conduct the interview.

tion, the structure of the question, the order in which the questions are presented, the overall topic of the questionnaire, whether the question is read by the respondent (self-administration) or is presented to the respondent by an interviewer, and the mode of communication used by the interviewer (that is, telephone versus face-to-face presentation). The wording of a question is often seen as one of the major problems in survey research: although one can standardize the language read by the respondent or the interviewer, standardization of the language does not imply standardization of the meaning. For example, "Do you own a car?" appears to be a simple question from the perspective of semantics and structure. However, several of the words in the question are subject to variation in interpretation, including "you" (just the respondent or the respondent and his or her family), "own" (completely paid for, purchased as opposed to rented), and even the word "car" (does this include vans and trucks?). The goal for the questionnaire designer is to develop questions that exhaust the range of possible interpretations, making sure that the particular concept of interest is the concept that the respondent has in mind when responding to the item.

One source of variation in a respondent's comprehension of survey questions is due to differences in the perceived intent or meaning of the question. Perceived intent can be shaped by the sponsorship of the survey, the overall topic of the questionnaire, or the environment more immediate to the question of interest, such as the context of the previous question or set of questions or the specific response options associated with the question.

Respondent as Source of Measurement Error

Once the respondent comprehends the question, he or she must retrieve the relevant information from memory, make a judgment as to whether the retrieved information matches the requested information, and communicate a response. Much of the measurement error literature has focused on the retrieval stage of the question-answering process, classifying the lack of reporting of an event as retrieval failure on the part of the respondent and comparing the characteristics of events that are reported with those that are not reported. Several factors have been found to be related to the quality of reporting, including the length of the reference period of interest and the salience of the information. For example, the literature suggests that the greater the length of the recall period, the greater the expected bias in the reporting of episodic information (e.g., Cannell et al., 1965; Sudman and Bradburn, 1973). Salience is hypothesized to affect the strength of the memory trace and, subsequently, the effort involved in retrieving the information from long-term memory.

The weaker the trace, the greater the effort needed to locate and retrieve the information.

As part of the communication of the response, the respondent must determine whether he or she wishes to reveal the information as part of the survey process. Survey instruments often ask questions about socially and personally sensitive topics. It is widely believed and well documented that such questions elicit patterns of underreporting (for socially undesirable behavior and attitudes), as well as overreporting (for socially desirable behaviors and attitudes). The determination of social desirability is a dynamic process and is a function of the topic of the question, the immediate social context, and the broader social environment at the time the question is asked. Even if the respondent is able to retrieve accurate information, he or she may choose to edit this information at the response formation stage as a means of reducing the costs associated with revealing the information.

The use of proxy reporters, that is, asking individuals within sampled households to provide information about other members of the household, is a design decision that is often framed as a trade-off among costs, sampling errors, and nonsampling errors. The use of proxy informants to collect information about all members of a household can increase the sample size (and hence reduce the sampling error) at a lower marginal data collection cost than increasing the number of households. The use of proxy respondents also facilitates the provision of information for those who would otherwise be lost to nonresponse because of an unwillingness or inability to participate in the survey interview. However, the cost associated with the use of proxy reporting may be an increase in the rate of errors of observation associated with poorer-quality reporting for others compared with the quality that would have been obtained under a rule of all self-response.

Most of the evaluations of the quality of proxy responses compared with the quality of self reports have focused on the reporting of autobiographical information (e.g., Mathiowetz and Groves, 1985; Moore, 1988) with some recent investigations examining the convergence of self and proxy reports of attitudes (Schwarz and Wellens, 1997). The literature is, however, for the most part silent with respect to the quality of proxy reports for personal characteristics, the exception being a small body of literature that addresses self-reporting versus proxy reporting effects in the reporting of race/ethnicity (Hahn et al., 1996) and the reporting of activities of daily living (e.g., Mathiowetz and Lair, 1994; Rodgers and Miller, 1997). The findings suggest that proxy reports of functional limitations tend to be higher than self-reports; the research is inconclusive as to whether the discrepancy is a function of overreporting on the part of proxy informants, underreporting on the part of self-respondents, or both.

Interviewers as Sources of Measurement Error

For interviewer-administered questionnaires, interviewers may affect the measurement processes in one of several ways, including:

- failure to read the question as written;
- variation in interviewer's ability to perform the other tasks associated with interviewing, for example, probing insufficient responses, selecting appropriate respondents, and recording the information provided by the respondent; and
- demographic and socioeconomic characteristics as well as voice characteristics that influence the behavior of the respondent and the responses provided by the respondent.

The first two factors contribute to measurement error from a cognitive or psycholinguistic perspective in that different respondents are exposed to different stimuli; thus, variation in responses is, in part, a function of the variation in stimuli. All three factors suggest that the interviewer effect contributes to an increase in variable error across interviewers. If all interviewers erred in the same direction (or their characteristics resulted in errors of the same direction and magnitude), interviewer bias would result. For the most part, the literature indicates that among well-trained interview staff, interviewer error contributes to the overall variance of estimates as opposed to resulting in biased estimates (Lyberg and Kasprzyk, 1991).

Other Essential Survey Conditions as Sources of Measurement Error

Any data collection effort involves decisions concerning the features that define the overall design of the survey, referred to here as the "essential survey conditions." In addition to the sample design and the wording of individual questions and response options, these decisions include the following:

- whether to use interviewers or to collect information via some form of self-administered questionnaire;
- the means for selecting and training interviewers (if applicable);
- the mode of data collection for interviewer administration (telephone versus face to face);
- the method of data collection (paper and pencil, computer assisted);
- whether to contact respondents for a single interview (cross-sectional design) or follow respondents over time (longitudinal or panel design);

- for longitudinal designs, the frequency and periodicity of measurement;
- the identification of the organization for whom the data are collected; and
- the identification of the data collection organization.

No single design feature is clearly superior with respect to overall data quality. For example, as noted above, interviewer variance is one source of variability that can be eliminated through the use of a self-administered questionnaire. However, the use of an interviewer may aid in the measurement process by providing the respondent with clarifying information or by probing insufficient responses. The use of a panel survey design, with repeated measurements with the same individuals, facilitates more efficient estimation of change over time (compared with the use of multiple cross-sectional samples); however, panel designs may be subject to higher rates of nonresponse (as a result of nonresponse at every round of data collection) or panel conditioning bias, an effect in which respondents alter their reporting behavior as a result of exposure to a set of questions during an earlier interview.

The following scenario is an illustration of statistical measures of error used by survey methodologists. Assume that the measure of interest is personal earnings among all adults in the United States. A "true value" exists if the construct of interest is carefully defined. The data will be collected as part of a household-based health survey being conducted by telephone. The decision to use the telephone for data collection implies that approximately 5 percent of the adults will not be eligible for selection. To the extent that the personal earnings of adults without telephones differ significantly from those with telephones, population-based estimates for the entire adult population will suffer from coverage bias. Similarly, not all eligible sample persons will participate in the interview because of refusal to cooperate, an inability on the part of the survey organization to contact the respondent, or other reasons, such as language barriers or poor health that limits participation. Once again, to the extent that the earnings of those who participate differ significantly from those who do not participate, population-based estimates of earnings will suffer from nonresponse bias.

If all respondents misreport their earnings, underreporting their earnings by 10 percent, and they consistently do so in response to repeated measures, the measure will be reliable but not valid and population estimates based on the question (e.g., population means) would be biased. However, multivariate model-based estimates that examine the relationship between earnings and human capital investment would not be biased, since all respondents erred in the same direction and relative magnitude. Differential response error, for example, the overreporting of earnings by low-income individuals and the underreporting of earnings by high-income individuals, may produce unbiased population estimates (e.g., mean earnings per person) but biased model-based estimates related to individual behavior.

MEASUREMENT ERROR: THE PSYCHOMETRIC PERSPECTIVE

The language and concepts of measurement error in psychometrics are different from the language and concepts used within the fields of survey methodology and statistics. The focus for psychometrics is on variable errors; from the perspective of classical true score theory, all questions produce unbiased estimates, but not necessarily valid estimates, of the construct of interest. The confusion arises in that both statistics and psychometrics use the terms *validity* and *reliability* to sometimes refer to very similar concepts and to sometimes refer to concepts that are quite different. Within psychometrics, the terms *validity* and *reliability* are used to describe two types of variable error. *Validity* refers to "the correlation between the true score and the respondent's answer over trials" (Groves, 1991, p. 8). The validity of a measure can be assessed only for the population, whereas the validity of both population estimates and individuals' responses presented in the survey methodological literature can be assessed.

Reliability refers to the ratio of the true score variance to the observed variance, where *variance* refers to variability over persons in the population and over trials within a person (Bohrnstedt, 1983). Once again, the measurement of reliability from this perspective does not facilitate measurement for a person but produces a measure of reliability specific to the particular set of individuals for whom the measurement was taken.

The psychometric literature identifies several means by which validity can be assessed; the choice of measures is, in part, a function of the purpose of the measurement. These measures of validity include content, construct, concurrent, predictive, and criterion. If one considers that the questions included in a particular instrument represent a sampling of all questions that could have been included to measure the construct of interest, content validity refers to the comprehensiveness as well as the relevance of those questions. Content validity refers to the extent to which the question or questions reflect the domain or domains reflected in the conceptual definition. Face validity refers to the extent to which each item appears to measure that which it purports to measure. Cognitive interviewing techniques that focus on the comprehension of items by respondents are, to some extent, a test of face validity.

Criterion-related validity evaluates the extent to which the measure of interest correlates highly with a "gold standard." The gold standard

222   *THE DYNAMICS OF DISABILITY*

could consist of a different self-reported measure, a behavioral measure, or an observation or evaluation outside the measurement process (e.g., clinical evaluation). Criterion-related validity is further categorized as concurrent validity or predictive validity. Concurrent validity refers to the correlation between the item of interest and some other item, event, or behavior measured at the same point in time, whereas predictive validity refers to the correlation between an indicator measured at time t and some other measure, event, or behavior measured at time $t + 1$.

When no gold standard exists, validity is evaluated in terms of the correlation between the measure of interest and other measures, according to theory-based hypotheses. As noted by McDowell and Newall (1996), "construct validation begins with a conceptual definition of the topic or construct to be measured, indicating the internal structure of its components and the theoretical relationship of scale scores to external criteria" (p. 33).

Measures of reliability include internal consistency (often referred to as coefficient Alpha or Cronbach's Alpha), test-retest, and interrater reliability. Internal consistency measures the extent to which all items in a scale measure the same underlying concept; it is only applicable for multi-item Likert scales. The reliability coefficient is a function of both the extent to which the items are homogeneous and the number of items in the scale; the coefficient increases with an increase in either the homogeneity of the items or the number of items. Test-retest reliability involves the measurement of the same person under the same measurement conditions at two points in time and can be used for single-item measures, as well as multi-item scales.[4] Interrater reliability refers to the consistency with which different raters or observers rating the same person agree with one another.

Returning to the example of the measurement of earnings to illustrate the measurement error properties of the construct in terms of psychometrics, assume that the question or questions designed to measure earnings are both comprehensive and relevant. Therefore, the questions would be assessed as having content validity (face validity). If, as noted above, all respondents underreported their earnings by 10 percent, the construct would have a lower score with respect to criterion validity, but since all respondents erred in the same direction and the same magnitude, the indicator would have construct validity. If repeated measurement resulted in consistent reports by all respondents, test-retest measures would indicate a high degree of reliability, not dissimilar to the conclusion drawn by statisticians.

---

[4]Within survey research, the conduct of a reinterview under the same essential survey conditions as the original interview is an example of a test-retest assessment of reliability.

POTENTIAL SOURCES OF MEASUREMENT ERROR SPECIFIC TO PERSONS WITH DISABILITIES

Similar to any other measurement of persons via the survey process, the identification of persons with disabilities is subject to the various sources of error discussed above. The measurement of persons with disabilities raises particular challenges, in light of the complexity of the phenomenon of interest and the demands of the measurement process. Some of the various sources that may be of particular importance are highlighted.

Coverage, Access, and Participation

The interactive nature of the survey interview places great demands on the sensory and physical resources of respondents. A face-to-face interview requires that the respondent have the capacity to hear the questions, respond orally, understand individual questions and response categories, and be able to maintain cognitive focus. In addition, the respondent must tolerate the physical demands of the interview, a task that may take up to an hour or two. Impairments or disabilities may limit a person's ability to participate in the survey process or limit access to the individual. The essential survey design features of a data collection effort can facilitate or limit access and participation of persons with disabilities. This is not unique to the measurement of persons with impairments or disabilities. The use of the telephone for data collection restricts the sample to those households with telephones; if the data collection by telephone does not accommodate the use of TTY technology, hearing-impaired individuals will also not be measured. Similarly, the use of self-administered paper and pencil questionnaires limits participation to those who are literate and whose vision permits the reading of the font size used on the questionnaire. The implementation of a self-response rule eliminates from measurement those for whom gatekeepers deny access and those, although they are willing to participate, who are unable to do so because of physical, mental, or emotional impairments or those for whom the barrier to participation is language, either their use of a different spoken language or their use of sign language.

Cognition and the Measurement of Persons with Disabilities

From a cognitive perspective, the measurement of persons with disabilities offers particular challenges. First, one needs to understand how individuals encode information about impairments and disabilities. In addition, effective questionnaire design requires an understanding of how the encoding of the information varies according to perceptual perspective (self-response versus other response, nature of the relationship between the respondent and the person for whom they are reporting). Second, little is known about how ability (capacity) is measured independent of environmental context (participation).

Many of the questions and sets of questions used to measure impairments and disability are plagued by comprehension problems related to both semantic and lexical complexity. For example, questions concerning work disability are subject to comprehension problems with respect to the shared meaning of "work." As noted earlier, the respondent must infer whether limitations in the kind or amount of work include factors related to transportation and access to the workplace. The desire for parsimonious means by which an individual's status can be assessed with respect to impairments or particular functional limitations has led to the creation of "composite" screening questions that nevertheless represent a single question and that may therefore be cost-effective, even though they press against the limits of working memory.[5]

The response task requires the respondent to retrieve information, determine the relevance of that information to the posed question, and formulate a response. Often the respondent is limited in the form of the response to a simple classification (e.g., "yes," limited in the kind or amount of work versus not limited) that fails to capture the full spectrum of the enablement-disablement process and the complexity of the phenomenon of interest. The mapping of this complex phenomenon to a limited number of response categories is most likely fraught with error.

The integration of theories of cognitive psychology with survey methodology has given rise to new methods of questionnaire design and evaluation. Many of the current measures of disability used in federal data collection efforts have not been subjected to testing methods common to new questions and questionnaires, for example, cognitive interviewing and behavior coding. Cognitive interviewing encompasses several techniques designed to elicit information about the respondent's comprehension of the question, the strategies by which the respondent attempts to retrieve information from memory, judgments as to whether the retrieved information meets the perceived goals of the question, and the formulation of responses. These techniques include the use of "think-aloud" protocols, follow-up probes, vignettes, and "sort-order" tasks (Forsyth and Lessler, 1991; Willis et al., 1991).

---

[5]For example: "Because of a physical, mental or emotional problem does anyone in the family have any difficulty with activities such as bathing, dressing, eating, getting in or out of a chair or bed, or walking across a room?"

A small body of literature has attempted to address problems in the comprehension of functional limitation questions in community-based survey interviews through the use of cognitive interviewing techniques (Jobe and Mingay, 1990; Keller et al., 1993). The findings from these investigations of functional limitation questions by use of cognitive interviewing techniques suggest that respondents varied in their interpretation of terms, tended to emphasize capacity rather than actual performance, overlooked qualifying statements within the question, failed to remember the use of human assistance, or failed to remember help with specific activities.[6]

Social Cognition, Self-Concept, and Social Desirability

What is meant when an individual is asked to classify him- or herself or someone else with respect to disability? Although reliable measurement may call for the use of clear, unambiguous, and objective definitions, it is questionable whether these goals are achievable with respect to the measurement of disability. Disability is a dynamic concept related to an underlying interface between an individual, societal accommodations and barriers, cultural norms and expectations, and behavioral norms. The use of "fuzzy logic" in which attributes apply only partially to given individuals may be more appropriate than standard survey techniques for the classification of disability (Hahn et al., 1996).

Although theories from cognitive psychology can provide information about the different cognitive processes by which self and proxy reporters engage in the response formulation process, one can turn to theories from social cognition to understand how individuals classify themselves and each other with respect to social categories. Although social cognition draws heavily from the theory and methods of cognitive psychology, as a subfield its focal point is on social objects, specifically, individuals or groups of individuals.

As noted by Brewer,

> In comparison to object categories, social categories have been postulated to be overlapping rather than hierarchically organized . . ., disjunctively rather than conjunctively defined . . . and more susceptible to accessibility effects. (Brewer, 1988, p. 1)

---

[6]See also Beatty and Davis (1998) for a cognitive evaluation of questions from the Survey of Income and Program Participation and the National Health Interview Survey concerning discrepancies in print reading disability statistics.

She further states that "social categories are assumed to be 'fuzzy sets' represented in the form of prototypical images rather than verbal trait lists" (Brewer, 1988, p. 10).

Social cognition also provides a theoretical perspective that provides information about divergent perspectives of actors and observers. The actor-observer difference suggests that actors draw on situational information to explain behavior at any given time, whereas observers use stable disposition properties of the actor to understand behavior (Jones and Nisbett, 1971). To the extent that proxy reporters view disabilities as stable as opposed to dynamic characteristics, one would anticipate discrepancies between self-reports and proxy reports.

Two sets of concepts drawn from social psychology are also useful for consideration with respect to the measurement of disability. The first is the concept of self; from a sociological perspective, self-conceptions involve three components: (1) how an individual sees him- or herself, (2) how other people actually see the individual, and (3) how the individual believes others see him or her (Rosenberg, 1990). The National Health Interview Survey-Disability Supplement (NHIS-D) and the National Organization on Disability/Harris Survey of Americans with Disabilities included questions that asked whether the respondent perceived that he or she had a disability and whether others perceived that the respondent had a disability. The second concept of interest involves the notion of social identity and the groups, statuses, and social categories to which the members of society are recognized as belonging. If the social identity category is ambiguous, the self-concept related to the social identity will also be ambiguous.

As noted by Jette and Badley in their paper, the measurement of disability is often presented in surveys as an "all or nothing phenomenon." This approach to the measurement of disability assumes that (1) the respondent recognizes and identifies with the socially defined label and (2) is willing to reveal membership in the group. If disability were an "all-or-nothing" phenomenon, identification with the classification would be less ambiguous; however, as already noted, the enablement-disablement process is a dynamic one, subject to variation as a function of both self and society. To the extent that identification or affiliation with group membership carries with it any type of social stigma, willingness to reveal membership in the group also carries with it a social cost, not unlike other phenomena subject to social desirability bias.

Ambiguous social classification categories are also more likely to be subject to context effects; respondents use the specific wording of questions, the immediately prior questions, or the overall focus of the question as a means for interpreting questions on disability. From a theoretical perspective, it is not surprising to find that estimates of the number of persons with disabilities vary as a function of differences in the specific wording of the question, the number of questions used to determine the prevalence and severity of impairments and disabilities, the context of the questions immediately proximate to the question of interest, and the overall focus of the questionnaire (health versus employment versus program participation).

EMPIRICAL EVIDENCE CONCERNING MEASUREMENT OF DISABILITY ERROR

To date, most investigations with respect to the error properties associated with the measurement of persons with disabilities or the measurement of persons with work disabilities have focused on errors of observation, ignoring differences in estimates due to coverage error and nonresponse error. This review of the empirical literature is therefore focused on errors of observation. As an illustration of the type of empirical investigations concerning error in the measurement of disability, this section begins by examining the work that has been done to date with respect to measures of activities of daily living (ADL). The intent is to provide an illustration of the type of work that has been done (and not done) with respect to a frequently used measure of functional limitation. The focus is then turned to the measurement of persons with work disabilities.

Measurement of ADLs, Functional Limitations, and Sensory Impairments

Although there are several different measurement methods for the assessment of physical disability, one of the most often used (within the context of survey measurement) is the Index of Activities of Daily Living, often referred to as the Index of ADL (Katz et al., 1963). The index was originally developed to measure the physical functioning of elderly and chronically ill patients, but several national surveys of the general population administer the index to adults of all ages. The index assesses independence in six activities: bathing, dressing, toileting, transferring from a bed or chair, continence, and feeding. Despite its wide acceptance and use, the psychometric properties of the index have not been well documented. Brorsson and Asberg (1984) reported reliability scores of 0.74 to 0.88 (based on 100 patients). Katz et al. (1970) applied the Index of ADLs as well as other indexes to a sample of patients discharged from hospitals for the chronically ill and reported correlations between the index and a mobility scale and between the index and a confinement measure of 0.50 and 0.39, respectively. Most assessments of the Index of ADLs have examined the predictive validity of the index with respect to independent living (e.g., Katz and Akpom, 1976) or the length of hospitalization and discharge to home or death (e.g., Ashberg, 1987). These studies indicate relatively high levels of predictive validity.

Despite the psychometric findings, a growing body of survey literature suggests that the measurement of functional limitations via the use of ADL scales is subject to substantial amounts of measurement error and that measurement error is a significant factor in the apparent improvement or decline in functional health observed in longitudinal data. Jette (1994) found that minor changes in the wording of the questions resulted in significant differences in the percentage of the population identified as being limited. Rodgers and Miller (1997) directly compared responses by the same respondents (or more specifically, for the same target individuals) by using different sets of ADL items and across different modes.[7] They conclude that the measurements of functional limitations with respect to counts of ADLs, indications of the use of assistive devices or personal help, and indications of any difficulty are all subject to large amounts of measurement error, of which a substantial portion is random error. Similar to other empirical work (e.g., Mathiowetz and Lair, 1994), their findings indicate that the use of proxy respondents results in higher levels of reporting, of which only 25 to 33 percent can be explained by demographic characteristics and health variables of the target individual. The finding suggests that higher levels of functional limitations reported by proxy respondents are not simply a result of selection bias, in which those with the most severe limitations are reported by proxy.[8] Their analyses also suggest that there was no clear effect of mode of data collection on estimates of functional limitations.

As illustrative of the variability and lack of reliability that is evident in survey estimates of functional limitations, Tables 1 and 2 present findings from the 1990 decennial census and the Content Reinterview Survey (CRS) (U.S. Bureau of the Census, 1993; McNeil, 1993). The CRS was conducted approximately 5 to 9 months following the 1990 decennial

---

[7]Note, however, that the allocation across modes was not experimentally varied but rather was an artifact in the design in which older respondents (80 years and older) were assigned to the face-to-face mode of data collection and those less than 80 years of age were assigned to the telephone mode of data collection. However, a substantial number of respondents were interviewed in the mode other than that to which they were originally assigned; the crossover permits determination of both main and interaction effects related to the mode of data collection.

[8]In comparisons of self-reports and proxy reports with clinical evaluations, Rubenstein et al. (1984) found self response to be more "optimistic" and responses obtained by proxy report to be more pessimistic, findings which suggest that both self and proxy responses are subject to measurement error, albeit in different directions.

TABLE 1 Mobility Limitations: Distributions to Census Question 19a and Content Reinterview Survey Question 34a, Persons 16 to 64 Years of Age, United States, 1990

Census Long Form: Difficulty Going Outside	Content Reinterview Survey: Difficulty Going Outside		
	Yes	No	Total
Yes	146	155	301
No	152	14,194	14,346
Total	298	14,349	14,647

NOTE: Prevalence rate based on Census: 2.03 percent, of which 49.0 percent were consistent responses. The prevalence rate based on the CRS: 2.05 percent, of which 48.5 percent were consistent responses.
SOURCE: McNeil, 1993.

TABLE 2 Self-Care Limitations: Distributions to Census Question 19b and Content Reinterview Survey Question 34b, Persons 16 to 64 Years of Age, United States, 1990

Census Long Form: Difficulty Taking Care of Personal Needs	Content Reinterview Survey: Difficulty Going Outside		
	Yes	No	Total
Yes	69	346	415
No	120	13,856	13,976
Total	189	14,202	14,391

NOTE: The prevalence rate based on census: 2.9 percent, of which 16.6 percent were consistent responses. The prevalence rate based on the Content Reinterview Survey: 1.3 percent, of which 36.5 percent were consistent responses.
SOURCE: McNeil, 1993.

census, with a sample of 15,000 housing units selected from among those housing units assigned to complete the long form of the census. With respect to mobility limitations, estimates from the two surveys appear to be similar (e.g., 2.03 versus 2.05 percent), but examination of the responses for individuals indicates a low rate of consistent responses (less than 50 percent) among those who reply affirmatively for either survey. With respect to personal care limitations, once again, a high rate of inconsistency in the responses is seen among individuals who respond affirmatively to the question in either survey. For example, among those 16 to 64 years of age, almost all (83.4 percent) of those who report a self-care limitation at the time of the census fail to report a self-care limitation in the CRS.

Comparison of the percentage of persons with mobility and self-care limitations from the two surveys is confounded by differences in the essential survey conditions under which the data were collected and that most likely contribute to the discrepancies evident in the data. These differences include:

- Differences in the mode of data collection. The decennial census is, for the most part, a self-administered questionnaire, whereas the CRS is interviewer administered and is conducted either by telephone (84 percent) or as a face-to-face interview (16 percent). McHorney et al. (1994) report that telephone administration of the SF-36 led to lower levels of reporting of chronic conditions and self-reports of poor health compared with a self-administered version of the SF-36.
- Differences in the context in which the questions were asked. Although the wording of the specific items is almost the same with respect to mobility limitations or self-care limitations, as can be seen from a comparison of the two questionnaires, the context in which the questions are asked differs in the two instruments. Several additional questions concerning sensory impairments, the use of assistive devices for mobility, mobility limitations related to walking a quarter mile or up a flight of steps, and the ability to lift and carry objects weighing up to 10 pounds precede the items of interest in the CRS. There is a large body of literature that documents the existence of context effects in attitude measurement (e.g., Schuman and Presser, 1981). The asking of additional questions could prime the respondent to think about impairments that he or she did not consider while answering the census questions, thereby resulting in an increase in the reporting of limitations. Alternatively, having just answered questions about a number of sensory impairments and limitations, respondents, when answering the more general questions, assume that the general question is intended to capture information not already reported; in this case one would expect the CRS estimates to be lower than those based on the census form. (See Sudman et al. [1996] for a review of the theoretical underpinning related to context effects and a thorough discussion of addition and subtraction effects.)
- Self-reporting versus proxy reporting. There is little information as to who provided information on either the census form or the CRS.

Although the CRS attempts to obtain self-reports from each adult household member, information for approximately 25 percent of the persons was reported by proxy. As noted earlier, proxy respondents tend to report more activity limitations and more severe limitations than self-respondents.

Finally, the possibility that the lack of reliability is indicative of the occurrence of real change between the time of the census and the time of the CRS must also be considered.

Although one can enumerate possible sources that explain the low rate of consistency between the two surveys, the lack of experimental design does not permit the identification of the relative contributions of the various design features to the overall lack of stability of these estimates.

Empirical evidence shows that even when questions are administered under the same essential survey conditions, responses are subject to a high rate of inconsistency. This evidence comes from the administration of the same topical module on functional limitations and disability to respondents in the 1992-1993 panel of the Survey of Income and Program Participation. The module was administered between October 1993 and January 1994 (Time 1) and then again between October 1994 and January 1995 (Time 2). The context of the questionnaire is the same in both waves; the topical module is preceded by the core interview, which focuses on earnings, transfer income, program participation, and other forms of income. Information is collected for all members of the household, usually by having one person report for himself or herself and all other family members. In addition, information as to who served as the respondent is recorded; thus one can examine consistency in the reporting of information across time among all self-responses. Table 3 presents selected comparisons of functional limitations and sensory impairments reported at Time 1 with those reported at Time 2. The comparisons clearly reveal high levels of theoretical inconsistency, even among self-respondents. For example, among those who report an inability to walk at Time 1, only 70.3 percent report the same status at Time 2. Limiting the comparison to self-reports only does not greatly improve the consistency. Among self-reporters, 76.7 percent of those reporting inability to walk at Time 1 report the same status in the subsequent interview.

These empirical findings illustrate some of the error properties associated with the measurement of functional limitations and sensory impairments. The research indicates that despite psychometric measures that indicate a relatively high degree of reliability, survey applications offer several examples of low levels of reliability, even under conditions in which the essential survey conditions are held constant. Subtle changes

TABLE 3 Selected Panel Survey of Income and Program Participation Data: Time 1 (October 1993-January 1994) and Time 2 (October 1994-January 1995) Comparisons, United States

Status at Time 1	All Cases		Self-Respondents Both Times	
	Number of Persons	Percent at Time 2 with Disability	Number of Persons	Percent at Time 2 with Disability
Uses cane, crutches, walker	508	45.5	286	50.0
Uses a wheelchair	175	61.7	83	68.7
Unable to see	159	49.1	87	49.4
Unable to hear	121	50.4	41	48.8
Unable to speak	47	68.1	5	80.0
Unable to walk	1,045	70.3	587	76.7
Unable to lift/carry	975	61.2	566	65.6
Unable to climb stairs	1,132	68.3	658	72.3
Needs help outside	699	53.5	302	57.3
Needs help bathing	271	52.0	114	54.4
Needs help dressing	237	49.8	80	55.0

SOURCE: McNeil, 1998.

in the wording of questions, the order of questions, or the immediate prior context offer further illustration of the lack of robustness of these items. Although one can enumerate all of the factors that *may* contribute to this volatility, the relative contributions of the various factors have not been experimentally determined.

Empirical Evidence Concerning Error in the Measurement of Work Disability

The assessment of work disability in federal surveys has focused on variants of a limited number of questions, most of which concern whether the individual is limited in the kind or amount of work he or she is able to do or is unable to work at all because of a physical, mental, or emotional problem. Not dissimilar to the assessment of functional limitations, work disability is measured in data collection efforts that vary with respect to the essential survey conditions, the specific wording of questions, the number of questions asked, and the determination of severity, duration, and the use of assistive devices or environmental barriers. As McNeil (1993) points out, one of the problems with the current set of indicators designed to measure work disability is that many fail to acknowledge the role of environmental barriers and accommodations. He states:

> Questions can be raised about the validity of data on persons who are "limited in kind or amount of work they can do" or are "prevented from working." The work disability questions make no mention of environmental factors, even though it is obvious that a person's ability to work cannot be meaningfully separated from his or her environment. Work may be difficult or impossible under one set of environmental factors but productive and rewarding under another. It would certainly be logical for a respondent to answer "no" to the question, "Do you have a condition that prevents you from working?" if the real reason he or she is not working is the inaccessibility of the transportation system or the lack of accommodations at the workplace. (pp. 3–4)

As noted in the paper by Jette and Badley, the "fundamental conceptual issue of concern is that health-related restriction in work participation may not be solely or even primarily related to the health condition . . .". One of the challenges facing questionnaire designers is the development of questions that match the conceptual framework of interest with respect to work disability, specifically, whether the focus is on the health condition that limits the individual's ability to perform specific tasks related to a specific job, the external factors related to the performance of work, other factors that affect participation in the work environment (e.g., transportation), or all three sets of factors.

Although McNeil (1993) raises questions concerning the validity of the work disability measures currently in use, several empirical investigations raise questions about the reliability of these measures, not unlike the findings with respect to the measurement of functional limitations and sensory impairments. Once again, it can be seen that differences in the wording of the questions, the context in which they are asked, the nature of the respondent, and other essential survey conditions, including the data collection organization and the sponsorship of the survey, may contribute to differences in estimates of the working-age disabled population.

Haber (1990), as revised from Haber and McNeil (1983), examined work disability from selected surveys between 1966 and 1988. He notes that "despite a high degree of consistency in the social and economic composition of the disabled population over a variety of studies, the overall level of disability prevalence has varied considerably" (p. 43). Haber's findings are reproduced in Table 4. The estimates from the various surveys represent differences in the year of administration, the wording of the questions, the overall content of the survey, the mode of administration, the organization collecting the information, and the organization

TABLE 4 Prevalence of Work Disability Across Various Surveys, United States, 1966-1982

	Percent Classified with a Work Disability		
Data Source (age range [years] for estimate)	Total	Males	Females
1966 SSA (18-64)	17.2	17.2	17.2
1967 SEO (17-64)	14.0	14.0	14.0
1969 NHIS (17-64)	11.9	13.1	10.9
1970 Census (16-64)	9.4	10.2	8.6
1972 SSA (20-64)	14.3	13.6	15.0
1976 SIE (18-64)	13.3	13.3	13.3
1978 SSA (18-64)	17.2	16.1	18.4
1980 Census (16-64)	8.5	9.0	8.0
1980 NHIS (17-64)	13.5	14.3	12.8
March, 1981 CPS (16-64)	9.0	9.5	8.5
March, 1982 CPS (16-64)	8.9	9.3	8.5
March, 1983 CPS (16-64)	8.7	9.0	8.3
March, 1984 CPS (16-64)	8.6	9.2	8.1
1984 SIPP (16-64)	12.1	11.7	12.4
March, 1985 CPS (16-64)	8.8	9.2	8.4
March, 1986 CPS (16-64)	8.8	9.4	8.2
1986 NHIS (18-64)	13.5	14.3	12.8

NOTES: SSA = Social Security Administration Disability Survey; SEO = Survey of Economic Opportunity; NHIS = National Health Interview Survey; SIE = Survey of Income and Education; March CPS = Annual March Supplement (Income Supplement) to the Current Population Survey; SIPP = Survey of Income and Program Participation.
SOURCE: Haber, 1990.

sponsoring the study. Although the wording of the questions is quite similar across the various surveys, there are some minor differences in specific wording (e.g., differences with respect to the emphasis on a health condition) and the order of the questions (e.g., whether the questions begin, as in the NHIS, by asking about whether a health condition keeps the person from working or begin, as in the SSA surveys, by asking whether the person's health limits the kind or amount of work that the person can do). As is evident from Table 4, the survey's content appears to be related to the overall estimate; the lowest rates of work disability prevalence come from the Census and the March Supplement to the Current Population Survey (8.5 to 9.4 percent), and the highest rates come from the surveys sponsored by SSA (14.3 to 17.2 percent).

The lack of stability that was evident for estimates of mobility and self-care limitations between the 1990 census and the CRS is also evident for estimates of work disability. Table 5 presents the comparison of

TABLE 5 Work Disability: Distributions to Census Questions 18a and 18b and Content Reinterview Survey Questions 33a and 33b for Persons 16-64 years of age, United States, 1990

Census Long Form: Limited in Kind or Amount of Work or Prevented from Working	Content Reinterview Survey: Limited in Kind or Amount of Work or Prevented from Working		
	Yes	No	Total
Yes	778	366	1,144
No	650	12,988	13,638
Total	1,428	13,354	14,782

NOTE: The prevalence rate based on census: 7.7 percent, of which 68 percent were consistent responses. The prevalence rate based on the Content Reinterview Survey: 9.7 percent, of which 54.5 percent were consistent responses.
SOURCE: McNeil, 1993.

responses between the 1990 census and the CRS with respect to whether the person is limited in the kind of work, or the amount of work, or is prevented from working at a job because of physical, mental, or other health conditions. Once again, it can be seen that between one-third and almost one-half of the respondents are inconsistent in their responses.

More recent investigations have used the extensive data from NHIS-D to investigate alternative estimates of the population with work disabilities. The data also provide an opportunity to examine inconsistencies in the reporting of work disability and receipt of SSI or SSDI benefits. For example, LaPlante (1999) found that, based on the data from the NHIS-D, 9.5 million adults 18 to 64 years of age report being unable to work because of a health problem. Among these 9.5 million adults, 5.3 million (or 56 percent) do not report receipt of SSI or SSDI benefits. If one looks at those who report receiving SSI or SSDI benefits, 75 percent report that they are unable to work and 13 percent report that they are limited in the kind or amount of work that they can perform, but 12.3 percent who report receipt of benefits do not report any limitation with respect to work.

Although these variations in estimates derived from different surveys suggest instability in the estimates of the proportion of persons with work disabilities as a function of the wording of the question, the nature of the respondent, and the essential survey conditions under which the measurement was taken, they provide little information about measurement error within the framework of either survey statistics or psychometrics. Little is known about the validity of these items or the reliability of these items, whether one views validity from the perspective of survey statistics as deviations from the true value or from the perspective of psychometrics as criterion-related or construct validity. The relative contributions of various sources of error are, for the most part, unknown; it is only known that various combinations of design features produce different estimates. None of the studies address errors of nonobservation.

QUESTION WORDING ISSUES RELATED TO SELECTED MEASURES OF WORK DISABILITY

Jette and Badley point out the conceptual problems inherent in many questions designed to measure persons with work disabilities, including the failure of most questions to enumerate the separate elements related to the role of work. That failure is evident in most work disability screening questions designed to be administered to the general adult population. The gap between the conceptual framework and the questions used to screen for work disability, is illustrated by using questions from several federal data collection efforts.

The long form of the decennial census for the year 2000 includes the following questions:

> Because of a physical, mental, or emotional condition lasting 6 months or more, does this person have any difficulty in doing any of the following activities: . . .
>
> d. (Answer if this person is 16 years old or over.) Working at a job or business?

The respondent is to check a box corresponding to "Yes" or "No."
The question is complex for several reasons:

- The respondent must consider multiple dimensions of health (physical, mental, and emotional) and attribute difficulty working at a job or business to one or more of these health problems. The explicit enumeration of physical, mental, or emotional conditions serves as a means of clarifying for the respondent the fact that the question is intended to cover all three dimensions of health, but at the cost of additional cognitive processing by the respondent.
- The respondent must also assess the duration of the condition and determine the degree to which the 6 months is intended to convey

6 months *specifically* or a more general concept of a "long-term" condition.

- The term "difficulty" is subject to interpretation. Cognitive evaluation of the term "difficulty" suggests that for some respondents the term implies capacity or ability to perform the activity but does not infer actual participation in the activity.
- What is or is not included in the concept of working is further subject to interpretation by the respondent (e.g., inclusion or exclusion of sheltered workshops).

As with many single screening items, the question fails to address accommodations that facilitate participation or barriers that prohibit participation. For example, if an individual is currently employed in an environment that accommodates a health condition, the respondent must determine whether the person should be considered as having difficulty working, even though the present employment situation presents no difficulty to the person.

The NHIS asks two questions concerning work limitations:

Does any impairment or health problem NOW keep _____ from working at a job or business?

Is ____ limited in the kind OR amount of work ___ can do because of any impairment or health problem?

In contrast to the questions in the census long form, the NHIS questions do not enumerate the various areas of health for consideration, nor does either question include a qualifying statement with respect to duration. The two questions are more specific in addressing the impact on working; compared with the term "difficulty" used in the census questionnaire, the NHIS probes whether a condition prevents the person from working or limits the kind or amount of work. Once again, note the lack of distinction between the ability to perform the activities associated with the actual performance of the job and those activities related to the role of work. For those who retire early because of a health condition or impairment, would the respondent consider that health problem as keeping the person from working?

IMPLICATIONS FOR METHODOLOGICAL RESEARCH

The point of the examples presented above is not to criticize the questionnaires in which they appear but rather to illustrate the problem of attempting to measure a complex, multidimensional, dynamic construct with a single question or a set of two questions. No one or even two questions can possibly tap into the various components of work disabilities. Clearly the first step toward a robust set of screening items is the acceptance of a shared conceptual framework and understanding of the dimensions of the construct of interest. That framework must consider the social environment in which the measurement of interest will be taken, understanding that the comprehension of the question is shaped not only by the specific words used in the question and the context of the question, but by the perceived intent of the question. The use of cognitive laboratory techniques can aid in the identification of problems of comprehension due to the use of inherently vague terms and differential perceptions of the intent of the question. Such techniques will aid in the understanding of the validity of the questions and, through the refinement of the wording of questions, hopefully improve the reliability of the items.

Simply documenting that variation in the essential survey conditions of the measurement process contributes to different estimates of persons with work disabilities is not sufficient; the marginal effects of various factors need to be measured and the impact needs to be reduced through the use of alternative design features. Both of these can be accomplished only through a program of experimentation. Similarly, the psychometric properties of these measures need to be assessed. Without undertaking a thorough program of development and evaluation, the discrepant estimates evident in the empirical literature will persist.

REFERENCES

Andersen R, Kasper J, Frankel M. 1979. *Total Survey Error.* San Francisco: Jossey-Bass Publishers.

Ashberg K. 1987. Disability as a predictor of outcome for the elderly in a department of internal medicine. *Scandinavian Journal of Social Medicine* 15:261–265.

Beatty P, Davis W. 1998. *Evaluating Discrepancies in Print Reading Disability Statistics through Cognitive Interviews.* Unpublished Memorandum. Washington, DC: U.S. Bureau of the Census.

Bohrnstedt G. (1983) Measurement. In Rossi, Wright, Anderson, eds. *Handbook of Survey Research.* New York: Academic Press.

Brewer M. 1988. A dual process model of impression formation. In: Srull T, Wyer R, eds. *Advances in Social Cognition, Volume 1.* Hillsdale, NJ: Lawrence Erlbaum Associates.

Brorsson B, Asberg K. 1984. Katz Index of Independence in ADL: Reliability and validity in short-term care. *Scandinavian Jour of Rehabilitation Medicine* 16:125–132.

Cannell C, Fisher G, Bakker T. 1965. Reporting of hospitalizations in the health interview survey. *Vital and Health Statistics*, Series 2, Number 6. Washington, DC: U.S. Public Health Service.

Forsyth B, Lessler J. 1991. Cognitive laboratory methods: A taxonomy. In: Biemer, Groves, Lyberg, Mathiowetz, Sudman, eds. *Measurement Errors in Surveys.* New York: John Wiley and Sons.

Groves R. 1989. *Survey Errors and Survey Costs*. New York: John Wiley and Sons.

Groves R. 1991. Measurement errors across the disciplines. In: Biemer P, Groves R, Lyberg L, Mathiowetz N, Sudman S, eds. *Measurement Errors in Surveys*. New York. John Wiley and Sons.

Groves R, Couper M. 1998. *Nonresponse in Household Surveys*. New York: John Wiley and Sons.

Haber L. 1967. Identifying the disabled: Concepts and methods in the measurement of disability. *Social Security Bulletin* 30:17–34.

Haber L. 1990. Issues in the definition of disability and the use of disability survey data. In: Levin, Zitter, Ingram, eds. *Disability Statistics: An Assessment*. Washington, DC: National Academy Press.

Haber L, McNeil J. 1983. *Methodological Questions in the Estimation of Disability Prevalence*. Unpublished report. Washington, DC: U.S. Bureau of the Census.

Hahn R, Truman B, Barker N. 1996. Identifying ancestry: The reliability of ancestral identification in the United States by self, proxy, interviewer and funeral director. *Epidemiology* 7:75–80.

Hansen M, Hurwitz W, Bershad M. 1961. Measurement errors in censuses and surveys. *Bulletin of the International Statistical Institute* 38:359–374.

Jette A. 1994. How measurement techniques influence estimates of disability in older populations. *Social Science and Medicine* 38:937–942.

Jette A, Badley E. 1999. Conceptual issues in the measurement of work disability. In: Mathiowetz N, Wunderlich GS, eds. *Survey Measurement of Work Disability: Summary of a Workshop*. Washington, DC: National Academy Press. Pp. 4–27.

Jobe J, Mingay D. 1990. Cognitive laboratory approach to designing questionnaires for surveys of the elderly. *Public Health Reports* 105:518–524.

Jones E, Nisbett R. 1971. *The Actor and the Observer: Divergent Perceptions of the Causes of Behavior*. Morristown, NJ: General Learning Press.

Katz S, Akpom C. 1976. Index of ADL. *Medical Care* 14:116–118.

Katz S, Ford A, Moskowitz R, Jacobsen B, Jaffe M. 1963. Studies of illness in the aged: The index of ADL: A standardized measure of biological and psychosocial function. *Journal of the American Medical Association* 185:914–919.

Katz S, Downs T, Cash H, Grotz R. 1970. Progress in development of the Index of ADL. *Gerontologist* 10:20–30.

Keller D, Kovar M, Jobe J, Branch L. 1993. Problems eliciting elders' reports of functional status. *Journal of Aging and Health* 5:306–318.

Kish L. 1965. *Survey Sampling*. New York: John Wiley and Sons.

LaPlante M. 1999. *Highlights from the National Health Interview Survey Disability Study*. Presentation to the Committee to Review the Social Security Administration's Disability Decision Process Research, Institute of Medicine, and Committee on National Statistics, National Research Council.

Lord F, Novick M. 1968. *Statistical Theories of Mental Test Scores*. Reading, Mass: Addison-Wesley.

Lyberg L, Kasprzyk D. 1991. Data collection methods and measurement error: An overview. In: Biemer, Groves, Lyberg, Mathiowetz, Sudman, eds. *Measurement Errors in Surveys*. New York: John Wiley and Sons.

Mathiowetz N, Groves R. 1985. The effects of respondent rules on health survey reports. *American Journal of Public Health* 75:639–644.

Mathiowetz N, Lair T. 1994. Getting better? Change or error in the measurement of functional limitations. *Journal of Economic and Social Measurement* 20:237–262.

McDowell I, Newall C. 1996. *Measuring Health. A Guide to Rating Scales and Questionnaires*. New York: Oxford University Press.

McHorney C, Kosinski M, Ware J. 1994. Comparisons of the costs and quality of norms for the SF-36 Health Survey collected by mail versus telephone interview: Results from a national survey. *Medical Care* 32:551–567.

McNeil J. 1993. *Census Bureau Data on Persons with Disabilities: New Results and Old Questions about Validity and Reliability.* Paper presented at the 1993 Annual Meeting of the Society for Disability Studies, Seattle, Washington.

McNeil J. 1998. Selected 92/93 Panel SIPP Data: Time 1 = Oct.93–Jan.94, Time 2 = Oct.94–Jan.95. Unpublished table.

Moore J. 1988. Self/proxy response status and survey response quality: A review of the literature. *Journal of Official Statistics* 4:155–172.

Rodgers W, Miller B. 1997. A comparative analysis of ADL questions in surveys of older people. *The Journals of Gerontology* 52B:21–36.

Rosenberg M. 1990. The self-concept: Social product and social force. In: Rosenberg M and Turner R, eds. *Social Psychology: Sociological Perspectives.* New Brunswick: Transaction Publishers.

Rubinstein L, Schaier C, Wieland G, Kane R. 1984. Systematic biases in functional status assessment of elderly adults: Effects of different data sources. *The Journal of Gerontology* 39(6):686–691.

Sampson A. 1997. Surveying individuals with disabilities. In Spencer B, ed. *Statistics and Public Policy.* Oxford: Clarendon Press.

Schuman H, Presser S. 1981. *Questions and Answers in Attitude Surveys.* New York: Academic Press.

Schwarz N, Wellens T. 1997. Cognitive dynamics of proxy responding: The diverging perspectives of actors and observers. *Journal of Official Statistics* 13:159–180.

Suchman L, Jordan B. 1990. Interactional troubles in face-to-face survey interviews. *Journal of the American Statistical Association* 85:232–241.

Sudman S, Bradburn N. 1973. Effects of time and memory factors on response in surveys. *Journal of the American Statistical Association* 68:805–815.

Sudman S, Bradburn N, Schwarz N. 1996. *Thinking About Answers: The Application of Cognitive Processes to Survey Methodology.* San Francisco: Jossey-Bass.

Tourangeau R. 1984. Cognitive sciences and survey methods. In: Jabine, Straf, Tanur, Tourangeau, eds. *Cognitive Aspects of Survey Methodology: Building a Bridge Between Disciplines.* Washington, DC: National Academy Press.

U.S. Bureau of the Census. 1993. *Content Reinterview Survey: Accuracy of Data for Selected Population and Housing Characteristics as Measured by Reinterview.* U.S. Department of Commerce, 1990 Census of Population and Housing, Evaluation and Research Reports. Washington, DC.

Willis G, Royston P, Bercini D. 1991. The use of verbal report methods in the development and testing of survey questionnaires. *Applied Cognitive Psychology* 5(3):251–267.

SSA's Disability Determination of Mental Impairments: A Review Toward an Agenda for Research

March 1999
(Updated October 2001)

Cille Kennedy, Ph.D.[1]

The Social Security Administration (SSA) operates two disability benefit programs; Social Security Disability Insurance (SSDI) for disabled workers and Supplemental Security Income (SSI) for disabled impoverished adults and children. Both of these programs come under periodic scrutiny. Of present concern is the process by which claims for disability benefits are adjudicated. In an effort to provide policymakers with a scientific base for future deliberations and indicated directions, the Institute of Medicine (IOM) and the Committee on National Statistics (CNSTAT) have been asked to examine the reliability, validity, adequacy, and appropriateness of SSA's current and proposed research activities as they related to the proposed redesign of the disability determination (Wunderlich and Kalsbeek, 1997).

The focus of this paper, commissioned by the Committee to Review the Social Security Administrations' Disability Decision Process Research (the committee), is on the determination of disability status of initial claims, based on mental disorders, for SSDI and SSI disability benefits. The scope of this paper covers the initial determination and emphasizes the medical aspects of the process. It is limited to the adjudication of adult claims. The paper draws heavily on the evaluation—contracted by the

---

[1]Cille Kennedy is a Policy Analyst at the Office of Disability, Aging and Long-Term Care Policy, Office of the Assistant Secretary for Planning and Evaluation, Department of Health and Human Services in Washington, D.C.

SSA and conducted by the American Psychiatric Association (APA)—of SSA's standards and guidelines used in the determination of claims based on mental disorders. Building upon this base, the paper reviews the conceptual model and taxonomy of the World Health Organization's (WHO) *International Classification of Functioning, Disability and Health* (ICF) (WHO, 2001) and related WHO disability assessment instruments, and their potential utility in the redesign of the determination process, and toward a cohesive agenda for research. The paper is intended to stimulate an agenda for research to inform future modifications of the disability determination whether or not a formal redesign is undertaken.

BACKGROUND

Statutory Definition of Disability

The foundation of the SSA's two disability programs is the statutory definition of disability in the Social Security Act. The same definition applies to both the SSDI and the SSI programs and is the standard that the SSA puts into operation for the determination of claims for disability benefits. The definition can be changed only by an act of Congress. According to the Social Security Act, the definition of disability is "Inability to engage in any substantial gainful activity by reason of any medically determinable physical or mental impairment which can be expected to result in death or which has lasted or can be expected to last for a continuous period on not less than 12 months" (Section 223(d)(1)(A)).

The term "substantial gainful activity" means work that is remunerated at a rate specified by regulations. As of January 1, 2001, the rate is $740 per month. In other words, an individual may not be earning more than $740 per month in order to be eligible for disability benefits. The statute further states that the physical or mental impairments must be so severe that the claimant cannot do *any* work in the national economy that exists in substantial numbers. It does not matter whether or not jobs are available in the local region or whether or not the person would actually be hired if a job existed. If a type of job exists in substantial numbers somewhere in the country that the claimant could do, then she or he is not given disability benefits.

Conversely, consideration *is* given to the person's age, educational level, and past work experience. A person nearing retirement age is treated differently by SSA than a younger, working age individual. The older person is more likely to be considered favorably for disability benefits. A person with a grade school education is not expected to be able to work at an available occupation that requires advanced educational expertise, and a person with a work history of manual labor is not expected to obtain an available position as a business executive. The bottom line for SSA disability is that the person cannot do the simplest, least demanding, existing work whether it is available or not.

The reason that an individual is unable to work must be due to a physical or mental impairment that is medically detectable. The statute goes on to define physical and mental impairments as resulting from "anatomical, physiological, or psychological abnormalities which are demonstrable by medically acceptable clinical and laboratory techniques" (Section 223(d)(3)). The effect of any one impairment or any combination of impairments is to be evaluated for severity in determining disability for work. If one impairment is so severe that the person cannot work, then the person is considered disabled. If a combination of impairments precludes work, then the person would be considered disabled even if no single impairment would be considered severe by itself (Section 223(d)(2)(C)).

Sequential Evaluation

The process by which SSA adjudicates initial claims for both SSDI and SSI disability benefits is called sequential evaluation. There are five steps in the initial process. Responsibility for completing these steps is divided between federal SSA workers in local SSA district offices (DOs) and state employees working in state Disability Determination Services (DDSs). The SSA contracts with state agencies, such as departments of social services or rehabilitation, to act as DDSs.

The *first step* in the sequential evaluation takes place at the local DO. The person claiming disability (or a representative) appears and applies for benefits. Here, the DO staff determines whether or not the individual is currently working according to the criteria established by the regulated amount considered to be "substantial" (currently $740 per month). If the person is earning at or above this level, the claim is denied at this step.

If the claim is not denied, then it is necessary to decide whether the claim should be processed for SSDI or SSI benefits, or both. Although the process is the same, the administrative criteria and cash award amounts differ. SSDI is an entitlement program for workers. There are specific work history requirements for SSDI. Workers pay into the Social Security system and therefore have the right to receive cash benefits if disabled. The SSI program is for people whose income and resources are below a certain monthly level and who are blind, aged, or disabled. Children are eligible for SSI. It is possible to receive both SSDI and SSI benefits simultaneously. Once the decision is made about which disability program the person is eligible for, the SSA DO staff requests that the applicant supply the necessary medical evidence and work history to support the claim of disability. The SSA DO staff then forwards the claim to the state DDS.

THE DYNAMICS OF DISABILITY

The *second step* begins when the application is received in the state DDS. Here a team collects and evaluates the medical evidence and work history. There must be sufficient medical evidence to substantiate a determination of the claimant's disability status. The team consists of a Disability Analyst and a Reviewing Medical Consultant. For claims based on mental impairments, the Reviewing Medical Consultant is usually a psychiatrist or clinical psychologist.

The Listings of Mental Impairments (the Listings) are the standard for the evaluation of the medical evidence for demonstrable signs, symptoms, and restrictions in daily life (described below). They are the translation of the medical component of SSA's definition of disability. The review of claims based on mental impairment is put into operation against the standards of the Listings using the Psychiatric Review Technique Form (PRTF) (described below.) The PRTF guides the decision-making process, conforming to SSA regulations (SSA Regulations, 404.1520 and 404.1520a) as to how decisions are to be made.

The Reviewing Medical Consultant determines whether an impairment exists, and—if so—whether the impairment is considered severe. An impairment for SSA is analogous to a diagnosis of one of nine alcohol, drug, or mental disorders. Severity is concluded on the basis of whether the claimant's condition results in slight or marked restriction of activities. If the impairment is found to be slight or 'not severe,' the claim is denied on the basis of this medical consideration alone. If the impairment is considered severe, the claim continues in the sequential evaluation process.

The *third step* involves severe cases only. The assessment at this step inquires whether the impairments are *so* severe as to preclude work on the basis of *medical evidence alone*. In this step, the Reviewing Medical Consultant decides whether or not the claimant's impairment(s) "meet or equal" standards set by the Listings of Mental Impairments.[2] If a case meets or equals the Listings as indicated on the PRTF, then the claimant is allowed benefits by this medical evaluation. If this severe case does not either meet or equal the Listings, it continues in the sequential evaluation.

Steps 2 and 3 are the only steps that permit a disability decision based on medical assessment alone. In step 2 the Reviewing Medical Consultant may determine that a claim is "not severe" and the claim is denied. The Listings of Mental Impairment are constructed in such a way that an

---

[2]A claim "meets" the Listings if the condition is of such severity that it precisely matches the findings. If the impairment does not match but is clinically equivalent to and exceeds the severity of one of the Listings (as is required in "meeting" a Listing), it is considered to "equal" the Listing that it most closely resembles.

individual who 'meets or equals' them in step 3 of sequential evaluation cannot reasonably be expected to work and the claimant is awarded SSA disability benefits.

The fourth step applies to those severe claims that have not been found so severe as to be allowed disability benefits on the basis of medical evidence alone. For these claims, the nonmedical factors of age, education, and work history are taken into account. Another difference in this step is that the Reviewing Medical Consultant provides additional input into the decision by completing the Mental Residual Functional Capacity Assessment (MRFCA) (see description below); it is the Disability Analyst who combines the narrative summary of the MRFCA with the age, educational level, and work history of the claimant to determine whether or not the claimant is capable of working at the level of her or his past employment. This decision is made in light of jobs available in the national economy. If the Disability Analyst finds that the claimant can do previous work, the claim is denied. If the finding is that the claimant cannot do previous work, the claim continues one final step in the initial review.

The fifth step applies the same claim material to the question of whether the claimant can do *any* job in the national economy. If the Disability Analyst determines that the claimant can do work—irrespective of whether it is locally available or whether the person would actually be hired—then the claim is denied. If the person cannot do any work in the national economy, then the claimant is awarded disability benefits.

The next section describes the standards and guidelines upon which these medical judgments are based.

Listings of Mental Impairments

The Listings of Impairments (SSA, 2001) are published by the SSA and updated periodically. Chapter 12 of the Listings is devoted to mental disorders, otherwise known as the Listings of Mental Impairments. Major revisions to the Listings of Mental Impairments, currently being applied to claims for disability benefits, were published in 1985[3] and have since undergone relatively minor modifications. The 1985 revision was intended to bring the Listings in line with then-current psychiatric practice to reflect the APA's third edition of its *Diagnostic and Statistical Manual of Mental Disorders* (DSM-III) (APA, 1980). The process of this revision was unique in SSA's history. It was the first time that the SSA had sought outside expertise to revise its Listings. The APA, the American Psychological

---

[3]The most recent edition of the Listings of Impairments was published in 2001. No substantive changes were made to the Listings of Mental Impairment that have impact on this report.

Association, and other mental health experts participated with the SSA in the process.

There are currently nine Listings of Mental Impairment designed to reflect the major psychiatric diagnostic categories likely to cause disability for work.[4] There are three sets of criteria within the Listings of Mental Impairments: A, B, and, for a subset of categories, C.

The purpose of the *A criteria* for mental disorders is to medically substantiate the presence of a mental disorder. The categories for adults follow:

- Organic Mental Disorders;
- Schizophrenic, Paranoid and Other Psychotic Disorders;
- Affective Disorders;
- Mental Retardation;
- Anxiety-Related Disorders;
- Somatoform Disorders;
- Personality Disorders;
- Substance Addiction Disorders; and
- Autistic Disorder and Other Pervasive Developmental Disorders.

The categories contain either two or three sets of criteria.[5] The A criteria are essentially diagnostic-like symptoms. They are not exact replicas of the DSM-III but are analogous to them. Each of the categories, except mental retardation and substance addiction, lists clinical findings. With the noted two exceptions, the threshold for the A criteria is that at least one of the clinical findings must be present. For example, to fulfill the A criteria for the category of Schizophrenia, Paranoid and Other Psychotic Disorders, there must be medically documented evidence of persistence, either continuous or intermittent, of at least one of the following:

1. delusions or hallucinations;
2. catatonic or other grossly disorganized behavior; or
3. incoherence, loosening of associations, illogical thinking, or poverty of content of speech associated with one of the following:

    - blunt affect,
    - flat affect, or
    - inappropriate affect; or

4. emotional withdrawal and/or isolation.

---

[4]The 1985 revision contained eight Listings of Mental Impairment.

[5]The Listing for Mental Retardation contains four sets of criteria. The Listing for Substance Addiction Disorders contains nine. They are outside the scope of this review.

The categories of Mental Retardation and Substance Addiction Disorders each differ and are not discussed as part of this review. They require special expertise and attention to detail beyond the scope of this paper.

The *B criteria* are applied to the category for which the A criteria are fulfilled. The purpose of the B criteria for mental disorders is to describe the functional restrictions that are incompatible with work and are associated with the mental impairments of the A criteria. The B criteria follow:

- marked restriction of activities of daily living;
- marked difficulties in maintaining social functioning;
- marked difficulties in maintaining concentration, persistence, or pace; and
- repeated episodes of decompensation, each of extended duration.

As an example, the SSA describes activities of daily living to include cleaning, shopping, cooking, taking public transportation, caring for one's grooming and hygiene, among others. These activities are referred to as activities of daily living and instrumental activities of daily living in the professional literature.

The *C criteria* apply to all categories except Mental Retardation, Personality Disorder, Substance Addiction, and Autistic Disorder.[6] The C criteria are additional considerations for cases that do not reach the threshold of the B criteria. For example, the C criteria for Schizophrenia, Paranoid, and Other Psychotic Disorders are intended to compensate for such instances when claimants are living in supportive residential settings or are otherwise adapted to a special environment that could not be sustained if the person went to work. These C criteria also consider individuals who have a history of serious episodes of disorder or disability and are currently functioning at a relatively high level through the benefits of medication but whose delicate functional status would be jeopardized by the additional stress of work. In other words, the relatively high degree of functioning is attributed to the compensatory medications or supports. Work would jeopardize this accomplishment and would vitiate the level of functioning attained. The C criteria for Anxiety-Related Disorders are designed to accommodate individuals with agoraphobia who are totally unable to function outside their homes but can function successfully within the home.

The following description of the forms show how decisions are made that put the sequential evaluation into effect for the medical component of the disability determination.

---

[6]Although the Listings for both Mental Retardation and Substance Addiction Disorders do have C criteria, they are of an entirely different conceptual nature.

Forms Used by the Reviewing Medical Consultant

The Reviewing Medical Consultant uses two forms in the sequential evaluation. The forms are used to document the existence of the medical condition and its impact on the domains of functioning related to the ability to work. The Psychiatric Review Technique Form puts the Listings of Mental Impairment into operation. The Mental Residual Functional Capacity Assessment is used to assess remaining functioning for claimants who are considered severe, but not sufficiently severe to be awarded benefits on the sole basis of the medical evidence using the PRTF. The MRFCA is intended to document what the person can do in spite of severe impairment. Unlike the B criteria of the Listings that ask the degree of limitation, the MRFCA intends to look at residual functioning—what the person can still do.

Psychiatric Review Technique Form This is designed to facilitate a review of the medical evidence and guide a medical decision as to the disability status of the claimant. The cover sheet, Section I, contains the summary of the medical review in two parts: the medical disposition and the category on which the medical disposition is based. The second page, Section II, provides space for the Reviewing Medical Consultant's notes. Following this is Section III, which lists the different categories of mental impairments along with their A criteria. At the top of each category are two checkboxes in which to indicate whether or not evidence of a cluster or syndrome exists that fits the particular diagnostic-like category. Beneath those two checkboxes, each of the category's A criteria is preceded by three checkboxes in which to indicate whether the specific item is present or absent, or whether insufficient evidence is provided in the medical information. The Reviewing Medical Consultant selects the one diagnostic-like category under which the claim will be reviewed and fills out the boxes for those items. If the A criteria are fulfilled, the Reviewing Medical Consultant then proceeds to the section that contains the four B criteria.

The page dedicated to the B criteria has two sections. The first is a chart that has the four B criteria, the areas of functional limitation listed on the left, and to the right, four or five checkboxes with which to rate the degree of limitation. Although degree of limitation is generally conceptualized as a continuum, for programmatic practicality, five intervals are identified. For example, restrictions in activities of daily living (the B1 criterion) can range from none, to slight, moderate, marked, and extreme.

These degrees of limitation are used to make two decisions: whether the claimant is (1) slightly limited (a step 2 denial) or (2) so severely limited that a benefit can be awarded (step 3). A slight impairment exists if all four B criteria are checked in the two left-hand columns (none, slight, never, or seldom). All other claims are considered severe. For a claim to be so severe as to meet or equal the Listings of Mental Impairments, two of the B criteria must be checked in the two columns to the right (marked, extreme, frequent, constant, repeated, or continual). As noted above, somatoform and personality disorders require that three of the B criteria must be so designated.

The C criteria are assessed as to their presence or absence.

This detailed description of the forms is not presented without reason. The judgments made by using the checkboxes result in two medically based disability determinations: denials for nonsevere claims and allowances for claims that are medically considered so severe as to preclude work. A "marked" limitation is considered a clinical decision. SSA describes "marked" as between a moderate and an extreme limitation. How sound are these decisions?

Mental Residual Functional Capacity Assessment This is used only for claims that are severe but have not been allowed disability benefits in the previous step of sequential evaluation, either meeting or equaling the listings. The MRFCA provides additional medical review for the Disability Analyst to combine with the nonmedical factors of age, education, and work history. The MRFCA is a checklist of 20 items subaggregated into four categories: (1) understanding and memory; (2) sustained concentration and persistence; (3) social interaction; and (4) adaptation. The form calls for a rating of limitation in the context of the individual's capacity to sustain the listed activity over a normal workday and workweek, on an ongoing basis. These items are rated on a three-point scale from "not significantly limited" to "markedly limited." Two other checkboxes permit ratings of "no evidence of limitation" and "not ratable on available evidence." "Not ratable" is to be used if the Reviewing Medical Consultant feels that there may be a limitation but cannot support a finding on the existing evidence. "No evidence" is for cases where none would be expected.

Of note is that item 11 essentially encapsulates the entire disability determination in one question: "The ability to complete a normal workday and workweek without interruptions from psychologically based symptoms and to perform at a consistent pace without an unreasonable number and length of rest periods." This is the essential question for the whole disability determination.

Using the 20 ratings as a foundation, the Reviewing Medical Consultant drafts a narrative in a section titled "Summary Conclusions." This narrative is the documentation that is used by the Disability Analyst. The ratings are not considered by the Disability Analyst.

Medical Evidence

The initial disability determination is a case record review, a paper review. It is the responsibility of the individual who is claiming to be disabled to provide the SSA with medical evidence to support the claim. Medical evidence consists of clinical signs, symptoms, and/or laboratory or psychological test findings. Clinical signs are medically demonstrable phenomena that reflect specific abnormalities of behavior, affect, thought, memory, orientation or contact with reality. A psychiatrist or psychologist generally assesses clinical signs. Symptoms are complaints presented by the individual. The findings may indicate an intermittent or persistent impairment depending on the nature of the disorder.

Medical evidence also includes information from other informed sources, such as family members and rehabilitation therapists, who have relevant knowledge of the claimant's functional capacity and limitations. This information is germane to the assessment of the B criteria on the PRTF and for the MRFCA review. There are no SSA-mandated forms for the provision of medical evidence. The collection of medical evidence is initiated by the local SSA district office and continued by the state-level DDS reviewing team to the point at which a disability determination of either an allowance or denial can be substantiated.

If the sources of medical evidence identified by the claimant do not provide sufficient evidence necessary to make a disability determination, a consultative examination can be provided. The SSA or DDS pays to have the claimant interviewed and a report sent to the DDS. The Consultative Examiner is generally someone not known to the claimant.

The Listings of Mental Impairment, and the forms used by Reviewing Medical Consultants—the PRTF and MRFCA—constitute the medical aspect of disability determination. For claims that do not result in a medical determination (i.e., a denial at step 2 using the PRTF because the disability is not severe or an allowance at step 3 because the disability is so severe that it precludes work on a medical review alone), the Disability Analyst continues the review with additional nonmedical factors. It is the medical aspect of the review of claims for disability benefits based on mental disorders that received a scientific evaluation.

AMERICAN PSYCHIATRIC ASSOCIATION EVALUATION STUDY

In 1984, prior to the publication of the 1985 Listings of Mental Impairments, the SSA, under the direction of the then-Assistant Commissioner for Disability Patricia Owens, contracted with the APA to design an evaluation of the soon-to-be-released standards and guidelines for the evaluation of mental impairments. The evaluation would include the Listings, the operational definitions for their implementation, and the forms (PRTF and MRFCA) that would be used in practice. The designed evaluation was accepted, and in 1985 the SSA awarded a contract to the APA to conduct the two-year evaluation.[7] The following description is a summary of the study.

Scope of the Study

The study was designed to ascertain the accuracy with which the medical standards and guidelines used by SSA's medical consultants operationalized the statutory definition of disability due to mental impairment and their consistency with contemporary psychiatric knowledge and practice. Within this broad objective, the study sought to identify and characterize cases that were difficult to evaluate with the medical standards and guidelines and to pinpoint the specific source of difficulty, and suggest solutions (Pincus et al., 1991).

Methodology

The study consisted of three components. The first investigated the compatibility of the SSA's revised medical standards and guidelines with the statutory definition of disability. This component provided the bulk of the study's empirical data. Component I employed 72 psychiatrists who were a demographically and professionally heterogeneous sample of APA's membership recruited from five geographically diverse cities. "Professional heterogeneity" meant the orientation or "school" of psychiatry represented, such as expertise in psychopharmacology or psychoanalysis, or experience with inpatient or outpatient, acute or chronic clients. Each psychiatrist was assigned to one of two study conditions. One—the sequential evaluation condition—received training and applied the SSA disability determination process and forms (e.g., PRTF and MRFCA) used by SSA's Reviewing Medical Consultants in actual casework. SSA staff participated in the training. The second study condition—the statutory definition condition—reviewed claims on the basis of psychiatrists' knowledge of the characteristics and limitations associated with the disorders experienced by the claimants. Training for this study condition consisted of in-depth discussions of the statutory definition of disability, claimants' impairments, functional limitations, and whether or not the claimants would be considered disabled according to the law. None of SSA's forms were used by this study condition.

---

[7]The Listings of Mental Impairments have been modified slightly since the APA conducted the evaluation. However, the essential study findings are still relevant.

Psychiatrists in both study conditions were instructed on the use of the Clinical Disability Severity Rating (CDSR), specially designed for the study. This rating was designed to record the psychiatrists' decisions as to the degree of disability for work experience by the claimants. This was necessary for two reasons. First, the statutory definition condition needed to document a decision as to disability status. Second, the sequential evaluation condition needed to record a disability finding at the fourth and fifth steps in sequential evaluation.[8] The CDSR is a 10-point scale from –5 (unable to work) to +5 (able to work) with no zero point, permitting a dichotomous rating with a negative score signifying inability to work. The CDSR was accompanied by a confidence rating that allowed the psychiatrists to scale the confidence of their CDSR judgments. Psychiatrists in the sequential evaluation condition scored claimants on the CDSR only after completing sequential evaluation and its forms.

The 36 psychiatrists in each study condition reviewed each of their assigned claims first as independent reviewers and then in a three-member panel within their study condition. Each panel was as demographically and professionally heterogeneous as possible. Each study condition had 12 panels. The rationale for the panels was that they would help participants make judgments reflecting the view of currently practicing psychiatrists.

The sample of initial claims for disability benefits based on mental disorders was provided by the SSA for the study. They had been filed by adults, had passed a quality assurance review by SSA regional offices, and were sanitized for confidentiality. One panel in both study conditions reviewed each claim.

The second study component was a survey of panelists to identify problems with the review process and to solicit suggestions about how to improve the review of claims.

The third study component was an in-depth narrative review of a subset of claims that had been rated differently by the two study conditions to understand the reasons for the discrepancy.

Major Study Findings

The overall study finding was that the proportion of agreement between panels of psychiatrists using SSA's process and forms and those using clinical judgment and additional discussion of the statutory definition of disability for each claim was 0.77, an agreement that was higher than chance (kappa of 0.46). The proportion of agreement and kappas

---

[8]The MRFCA does not lead to a medical disability decision. The MRFCA narrative must be combined by the nonmedical Disability Analyst with age, education, and work history.

within study conditions and at the individual level were similar and led to the conclusion that differences could be attributed to inherent difficulty of the claim, complexity of the sequential evaluation process, or the medical standards and guidelines themselves. Analyses further explored these sources of disagreement.

An analysis examined the agreement between study conditions when difficult claims were eliminated and found that the proportion of agreement between study conditions was 0.96 (kappa of 0.78), suggesting that problems in evaluation can be attributed largely to the inherent difficulty of claims. If the standards and guidelines led to markedly different disability judgments than the statute, then a far greater degree of disagreement would have been found between the two study conditions. Nonetheless, the findings do not rule out the possibility that improvements could be made to the standards and guidelines that would result in less difficulty and disagreement about disability status in the determination process.

Analysis of the Listings/PRTF

The next series of analyses examined the agreement between study conditions for each of the seven categories of mental impairment of the Listings as reviewed using the PRTF. Three of the Listings categories— Organic Mental Disorders; Schizophrenia, Paranoid and Other Psychotic Disorders; and Anxiety Disorders—were found to work well. Affective Disorders showed a statistically greater chance of disagreement about disability decisions than other disorders. Personality Disorders and Mental Retardation and Autism[9] had low agreement rates but were not statistically significantly. The sample size for these two disorders was small. The sample size for Somatoform Disorders was too small to interpret its high disagreement rate properly.

Within the Listings, analyses explored the agreement for use of the *A, B, and C criteria*. Moderately high rates of agreement on the selection of which A criteria to adjudicate a claim were found for Organic; Schizophrenic, Paranoid and Other Psychotic; Affective; Mental Retardation and Autism; and Anxiety Disorders. There was less concurrence in panelists' selection of Somatoform Disorders and Personality Disorder as Listing categories in which to adjudicate a claim. Ratings of agreement for the B criteria were reasonably high in the aggregate, but less reliable for the individual B criteria. Additional analysis found the first three B criteria (activities of daily living, social functioning and deficiencies in concentra-

---

[9]Mental Retardation and Autism were combined as one listing in 1985.

tion, persistence, and pace) to be valid whereas the fourth B criterion (episodes of deterioration or decompensation in a work-like setting) showed some reliability and validity but was weak overall.[10]

The C criteria for schizophrenia, paranoid, and other psychotic disorders were found to have low reliability but were valid. There were insufficient data to analyze the C criteria for Anxiety-Related Disorders. [11]

In Component III of the study, panelists reported that the period of review for the C criteria was confusing. It required a rating period of two or more years. Additionally, they noted that a large amount of inference was needed in rating the C criteria. For example, there is often no direct evidence of how claimants would function outside a highly supportive living situation if they have lived there two years or more. The clinical inference required to make this judgment as if the claimant did not live there was difficult.

MRFCA

Only the 20-item worksheet of the MRFCA was amenable to statistical analysis. Interrater reliability of the items was conducted. There were sufficient data to conduct two analyses. The first was whether there was any limitation for the given item (rating categories: "not significantly limited," "moderately limited," or "markedly limited" versus rating "no evidence of limitation in this category" or "not ratable on available evidence"). The second was the agreement among raters about the degree of limitation when a limitation existed. Reliability ratings were—at best—moderate. For some items, the reliability was so low as to suggest dropping the item. For most items, the reliability of degree of limitation was greater than deciding whether or not a limitation existed. This finding suggests that raters need further guidance in determining whether sufficient evidence is present.

Medical Evidence

During the independent reviews, each panelist rated each claim on the overall quality of the medical evidence in the claim folder on a five-point scale from "completely adequate" to "very inadequate." The average rating fell between "mostly adequate" and "somewhat inadequate." This finding must be taken in the context of the selection of claims for the study. They were not routine claims; SSA had selected them from a subset

---

[10]The phrasing of the B criteria has been modified since 1985.
[11]In 1985, these were the only two listings to have C criteria.

of claims based on their having passed a quality assurance for development of the medical evidence.

Although the survey of panelists in Component II of the study generally supported all of the above findings, panelists spontaneously reported that one of the major difficulties they encountered in the review of claims was: the lack of sufficient medical and other evidence; and the variability in the quality of medical evidence when present. They also noted that the chronological organization of the medical evidence based on the date of its receipt created difficulties in the review of claim folders.

Date of Onset and Period of Review

In Component III, panelists raised the issue of the difficulty in formally establishing the date of onset of the impairment as opposed and in addition to the date of onset of the disability. There is no place to document the clinician's decisions about the period under review on the PRTF. Without this documentation, psychiatrists discovered that they were rating different periods, sometimes leading to differences in clinical opinion, which resulted in consistent judgments when the timeframe was aligned.

Identification of Difficult Claim Types

For the purpose of allocating sources of discrepancies in order to home in on problems with the standards and guidelines, difficulties in the review process itself and difficult claim types were also examined. Difficult claims were those for which panel members disagreed about disability status. Analysis of difficult claim types was conducted in two ways. The first was to use study variables (e.g., panelist ratings of their confidence in their decisions and their ratings of quality of the medical evidence). These variables were used for the study's purpose.

The second approach was designed to identify factors that SSA could use on routine cases to predetermine which cases were amenable to a facile review and which might require more concentrated effort—perhaps a tiered review process: one process for "easy" claims and another for "difficult" claims.

The second approach was to use clinical and demographic variables found in routine SSA disability claims folders that would predict difficulty in the adjudication. The "clinical" factors that were included in the analysis were the presence of a notable physical disorder; the time of onset of impairment; work history; and notable alcohol or drug abuse. The demographic variables were age and gender. Three factors predicted difficulty regardless of the category of Listings/PRTF under which the

claim was reviewed or the step of sequential evaluation at which disability status was determined:

1. presence of a notable physical disorder amongst claims based on
   mental impairment;
2. onset of impairment less than 12 months prior to review; and
3. being female.

In the SSA system, these claims would likely take extra time to review
and be those—if denied benefits—that might be reversed on appeal.

Selected APA Recommendations

Based on study findings, recommendations were made in reference to
areas in the standards and guidelines that warranted potential revision
and to additional research that would further enlighten SSA's disability
determination. *All recommendations were made on the premise that the
basic construct of the SSA's medical standards and guidelines for the
evaluation of claims based on mental impairment should be retained.*
Recommendations from the APA study, described below, reflect those
that are consonant with the focus of the IOM committee and are detailed
to indicate areas for a research agenda.

First, the medical standards and guidelines for claims based on mental
impairment can be improved by refinements of specific aspects. In particular, the study identified the following eight areas:

1. Listing/PRTF category for Affective Disorders;
2. the A criteria for Affective and Personality Disorders;
3. the B4 criterion (episodes of deterioration or decompensation in
   work or work-like setting that cause the individual to withdraw
   from the situation or experience exacerbation of signs and symptoms);
4. scale points for all four B criteria;
5. timeframe and instructions for C criteria;
6. extending C criteria to other episodic disorders (e.g., Affective Disorders);
7. onset of impairment and period of review; and
8. MRFCA items.

Second, the medical evidence upon which disability decisions are
based needs improvement. A nationally standard form to be used by
treating psychiatrists, psychologists, consultative examiners, and other
clinicians should be designed and tested for collection of medical evi-

dence for all claims based on mental impairment. Third, identification and special evaluation of difficult claims should receive consideration within the SSA system. Fourth, a systematic series of studies and research development activities in collaboration with other federal agencies, academic institutions and professional organizations should be developed and conducted. Three areas of endeavor were identified:

1. SSA should conduct a review of the study claims to identify whether or not the same medical standards and guidelines lead to different decisions within all levels of SSA adjudication.
2. A longitudinal study of the claims reviewed in the study should be considered to understand the natural course of those claims in the SSA system and the course of the disability status of those claimants.
3. SSA should investigate the use of panels to adjudicate difficult claims.

The APA provided additional detail about these recommendations and described other recommendations outside the scope of the IOM committee's interest. It is possible that SSA has conducted some of this work.

Because the premise upon which all APA recommendations were made was that the basic construct of SSA's medical standards and guidelines should be maintained for claims based on mental impairment, any changes, modifications, or refinements would have to be based on input that is conceptually compatible and scientifically robust. The WHO's ICF is such a resource. The next section describes the ICF in general and then as it specifically pertains to the SSA disability determination.

INTERNATIONAL CLASSIFICATION OF FUNCTIONING, DISABILITY AND HEALTH

The ICF is one of the "family" of WHO classifications designed for use in a range of health and health-related applications. The classifications cover a broad range of health information issues and provide an international, standard language that enables communication throughout the world among various disciplines and sciences. For example, the WHO's *International Statistical Classification of Diseases and Related Health Problems, Tenth Revision* (ICD-10, or herein referred to as ICD) (WHO, 1993) is one of the most familiar. The ICD is a classification used to translate diagnoses of diseases and other health problems into an alphanumeric code for recording morbidity and mortality. The ICD has become the international standard diagnostic classification for all general epidemiological and many health management purposes. ICD and ICF complement each other.

The ICF classification covers a person's functioning and disabilities associated with health conditions at the body, person, and society levels. Functioning in the ICF refers to all body functions, activities, and participation as an umbrella term. Similarly, disability describes the impairments of body function, limitations in activities, and restrictions in participation. For activities and participation, it can capture both capacity and performance. The overarching aim of ICF is to provide a unified and standard language to serve as a scientifically based frame of reference for the description of health and health-related states. It is designed for use with physical and mental health conditions alike. Additionally, it specifically states that social security programs are among its potential applications. It will also provide the basis for a systematic coding scheme for health information systems that permits comparison of data across countries, health care disciplines, service settings, and time—data that complement ICD.

At a global level, three features of ICF compel further consideration by SSA. First, ICF is predicated on a universal model. The ICF is intended for use with all people, not a predetermined set identified as "the disabled." It is designed in such a way that different program criteria and thresholds can be applied to it: it does not contain a standard criterion of its own.

Second, ICF was developed with an internationally cultural perspective that has direct application to the heterogeneous population within the United States. The ICF was created with cross-cultural application as part of the process. It was not developed solely in the industrialized North American and Western European cultures. Thus, ICF has applicability to the diverse populations in the United States.

Third, one component of the WHO family of classifications is the development of specialty-based adaptations. For example, the ICD-10 has a version for use in primary care settings. With sufficient interest, a special adaptation of ICF for work disability could be developed. The factors associated with work could be compiled, leaving the relative weights of associated factors and the threshold for any dichotomous (able or not able to work) or interval (percent of work disability) decisions to SSA's program criteria to operationalize.

The ICF was unanimously endorsed at the May 2001 World Health Assembly in Geneva by the Ministers of Health of its 190 member countries. The ICF was published in all six official WHO languages and launched in October 2001. Because of its recent publication, attention to its predecessor is warranted.

International Classification of Impairments, Disabilities, and Handicaps (ICIDH-1980)

In the late 1970s there was a groundswell of interest in the impact of ill health on a person's functional status particularly as it related to provision of appropriate health care. Advances in medical knowledge were preventing mortality from acute medical conditions. This resulted in more people living longer, but living with chronic health conditions and ensuing disabilities. The classification scheme of the first edition of the WHO *International Classification of Impairments, Disabilities, and Handicaps: A Manual of Classification Relating to the Consequences of Disease* (ICIDH-1980) (WHO, 1980) was created to offer a framework to facilitate the provision of health-related information notably for chronic, progressive, or irreversible diseases. The ICIDH-1980 posited a model of the sequence underlying the illness-related phenomena as disease leading to impairment, leading to disability, and on to handicap. ICIDH-1980 defined these three consequences of the health condition. *Impairment* is defined as any loss or abnormality of psychological, physiological, or anatomical structure or function; *disability* as any restriction or lack (resulting from an impairment) of ability to perform an activity in the manner or within the range considered normal for a human being; and *handicap* as a disadvantage for a given individual, resulting from an impairment or a disability, that limits or prevents the fulfillment of a role that is normal (depending on age, sex, and social and cultural factors) for that individual.

This landmark work was not without controversy. When juxtaposed against other disability models, such as that proposed by Saad Nagi, inconsistencies and irregularities became evident. This too was the finding of an IOM committee (Pope and Tarlov, 1991).[12] However, ICIDH-1980 was the only model with an associated taxonomy.

ICIDH-1980 was accepted in many countries throughout the world. It is widely accepted and used in physical rehabilitation in The Netherlands. In France, the ICIDH-1980 definition of "handicap" became the basis for the national law upon which its disability benefits are based. Some research instruments based on ICIDH-1980 were developed, and a nascent body of disability research exists. The ICIDH-1980 was not generally accepted within the United States.

After a decade of use in services and research, the need for a revision became apparent. In the early 1990s the WHO undertook the revision of ICIDH-1980.

---

[12]Because of the thorough review by this other IOM committee, the presentation of ICIDH-1980 here is brief.

ICF Revision Process

The ICF revision process has four unique features intended to address identified shortcomings of ICIDH-1980 and to make it useful and applicable to the disablement experience in a range of relevant situations. First, ICF was developed as an iterative process. Initially, WHO worked with three collaborating centers and one international task force, the International Task Force on Mental Health, and Addictive, Behavioural, Cognitive, and Developmental (MH/ABCD) Aspects of ICIDH,[13] to draft an Alpha version (with the acronym of ICIDH-2 at that time).

With this version, the WHO began to build up the second unique feature of ICF: to include input in the development of ICF from diverse cultures, languages, and geographical areas. The Alpha version was circulated for review and comment, and new collaborating centers were incorporated into the process. This aspect of the revision process breaks from the tradition of developing a classification in English, based on North American and Western European expertise and then translating it into different languages. The ensuing Beta-1 version (also with the acronym ICIDH-2), which incorporated feedback from the Alpha draft, was even more widely distributed internationally.

The third unique feature was a series of formal field trials that were designed to collect empirical evidence for additional revision. Included in the field trials were such queries as the cross-cultural applicability of the concepts and model underlying ICF and the meaningfulness and sensitivity of the ICF domains and items in different cultures throughout the world. Although field trials are not unique to the development of international classification systems (e.g., the ICD), they are unique to the ICF process. Beta-1 field trials obtained data from all major populated geographical areas and continents: North and South America, Europe, Turkey, Russia, India, Japan, Africa, and Australia. Beta-2 field trials were even more extensive.

The dissemination of the field trials highlights the fourth unique feature of ICF. The ICF is predicated on the biopsychosocial model that combines the best of both the medical and the social models related to the WHO definition of health. Participants in the Beta-1 and Beta-2 field trials included not only professional mental and physical care providers, but also administrators, advocates, family members, and people with disabilities themselves. The SSA participated in the Beta-2 field trials. Unlike

---

[13]The SSA had been a formal member of the International MH/ABCD Task Force since its inception. Upon endorsement of the ICF by the World Health Assembly, all task forces were disbanded.

ICIDH-1980, ICF is intended for use by the broadest of audiences in the array of cultures, professions, and people affected by disabilities. As noted above, the U.S. Secretary for Health and Human Services, along with other Ministers of Health, endorsed the ICF as an official member of the WHO family of classifications at the World Health Assembly in May 2001.

Conceptual Model

In response to feedback, the ICF has been written in neutral rather than negative terminology. Advocates and people with disabilities in particular pointed out that the negative perspective of ICIDH-1980 and its terminology often were perceived as a negative description of the person with the disability rather than as a descriptive term about the disabling situation itself. Thus, ICF has employed neutral terms to the extent possible.[14] As with the ICIDH-2, the ICF envisions three components: body functions and structure, activities, and participation. Body functions are the physiological functions of body systems including psychological functions. Body structures are anatomical parts of the body such as organs, limbs, and their components. An activity is the execution of a task or action by an individual. Participation is involvement in a life situation. Decrements or disabilities in these three domains are respectively considered impairments, activity limitations, and participation restrictions. Impairments are problems in body function or structure as a significant deviation or loss. Activity limitations are difficulties that an individual may have in executing activities. Participation restrictions are problems an individual may experience in involvement in life situations.

Contextual factors are a significant addition to the conceptual formulation of ICF. Contextual factors include both personal and environmental factors. Environmental factors are the physical, social, and attitudinal backgrounds of the person. They are external to the individual and may have a positive or negative impact on the person's functioning. Inclusion of the list of environmental factors is an innovation of the ICF. Personal factors, such as gender, age, and level of education, also have an important impact on the person's functioning but are not listed in the ICF.

The conceptual model of ICF is an interactive one. ICIDH-1980 was linear. Not only does the health condition have an effect on the function and structure of the body, on the limitation of a person's performance of

---

[14]For example, "work" is a neutral term. ICF retains that neutrality by classifying it under the component of "activity" (also a neutral term) but uses a negative rating of difficulty to assess limitations in the performance of activities.

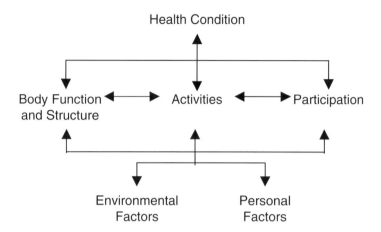

FIGURE 1. Interaction between the components of ICF.

activities, and on participation in society, but these three components of functioning can have a multidirectional impact on each other and on the health condition as well (Figure 1). Prior to the Americans with Disabilities Act of 1990 (ADA) people with ill health or with a disability were often limited in their participation in society. For example, an individual with a mobility disability might not have been hired for a job for which she or he was otherwise qualified. In this example, the ADA can be understood as an environmental factor that facilitates the reduction of barriers to participation in work.

The ICF is also seen as a dynamic, not a static, model. The components, domains, and items may be impaired, limited, or restricted either temporarily or permanently. Impairments, limitations, and restriction may be progressive, regressive, or static, and intermittent or continuous. The temporal quality is not fixed by ICF. The deviation from the norm may be slight or severe and may fluctuate in degree over time. ICF does not specify a threshold for the degree of deviation from the norm for a determination of disability status. It can be applied to legal definitions or program criteria and used to put them into operation.

The SSA's statutory definition of disability, some sequential evaluation components, and even the basic premise of SSA's disability benefits can readily be mapped onto this interactive conceptual model. For example, in SSA's definition of disability there must be a medically determinable physical or mental impairment that causes the inability to work. SSA's requirement for the impairments is that they either result in death or last at least one year; this is consistent with the ICF model. Additionally, the threshold for disability status can be accommodated by the model as well: it is designed to be applied to Social Security disability programs. In sequential evaluation, SSA considers the existence of jobs in the national economy; this is an environmental factor. Personal factors include gender, age, and education. SSA considers age, educational level, and work history. Disability benefits are essentially the replacement of earned income (stemming from SSDI) that facilitates participation in various domains of social life. None of these aspects of SSA disability program are inconsistent with ICF. It is intended for application to programmatic requirements.

Taxonomy

WHO uses an image of tree, branch, twig, stem, and leaf to describe the taxonomic structure. The entire classification of functioning and disability is the tree, components are branches, domains are twigs, items are stems, and subitems are leaves. Although the core concepts underlying the three components (i.e., body functions and structures, activities, and participation) are distinct, when applied to daily life the boundary between the person's capacity and performance, and the environmental impact upon the expression of activities and participation, are virtually impossible to depict without being artificial or without an evidence base. Thus, two of the ICF's components are contained in one list of the taxonomy: activities and participation. Conversely, as noted above, the components of body functions and structure are separately classified.

The ICF is organized in one, two, and full levels of detail. Additionally, it contains an index. For the uninitiated, using the one- or two-level classification serves to familiarize the organization of the material and assists in locating relevant items. The full classification contains the definitions, inclusions, and exclusions.[15]

The operational definition permits ready understanding of the item for laypeople, for clinicians, and for researchers alike. The definitions apply across cultures and avoid trendy jargon, particular schools of thought, or specific professions.

Coding and Rating

To use this neutrally termed classification and to identify impairments, activity limitations, or restrictions in participation, each compo-

---

[15]The ICF is published in two versions. The short version contains the one- and two-level classification with the second level having definitions.

nent of the ICF is accompanied by instructions for rating. Like the ICD, the ICF version has an alphanumeric coding system. The component, domain, item, and subitem are identified to the left of a decimal. To the right of the decimal, the coding structure contains space for qualifiers such as difficulty in performing activities or in being involved in life situations.

Impairments in ICF body functions are rated by one qualifier as the extent of the impairment on a five-point scale from "no impairment" (0) to "complete" (4), with mild, moderate and severe being 1, 2, and 3, respectively. This is generally thought of as a clinical rating, with the threshold determined by clinical expertise. Body structures have three qualifiers. The first is identical to body functions. The second qualifier uses nominal coding to describe the nature of the impairment. For example, 1 is used to indicate that there is a total absence of the body structure as in an amputated limb, 4 indicates aberrant dimensions of the body structure, and 6 denotes a deviation in the position. The third qualifier for body structure identifies the location of the impairment. For instance, 0 specifies that the impairment is in more than one region, 1 that it is on the right, 4 that it is in the front, and 7 that it is distal. Using all three qualifiers provides a rich description of the extent, nature, and location of the impairment of body structure.

Activities and participation have two qualifiers: performance and capacity. The first qualifier is performance, which demarks how much difficulty the person experiences in executing the task or being involved in a life situation as conducted in her or his usual day-to-day life. This means how and where the person spends time doing things. The second qualifier is capacity, which describes the person's ability to execute a task or action. Capacity is assessed in a formal test setting that is standardized to evaluate the true ability without the influence of the environment. Thus, capacity is considered to be evaluated in a neutral, standardized, or uniform environment. Variations of these two qualifiers indicate whether or not the person is using assistive devices or personal assistance in the assessment. Coding for both performance and capacity qualifiers is identical to body functions and to the first qualifier of body structures: it identifies the degree of difficulty.

In the structure of ICF qualifiers, there is room for consideration of additional qualifiers. It is anticipated that some users of ICF may develop their own qualifiers to suit their programmatic or other needs. For example, the SSA may wish to consider application of these or other qualifiers—or the development of an explicit algorithm using qualifiers—in refinement of the B criteria or revision of MRFCA.

Disability for Work

At the beginning of the ICF revision process, the International MH/ ABCD Task Force had a unique role. It was the only entity with responsibility for input into all aspects of ICF. Initially, each of three collaborating centers had responsibility for the first draft of either the then-labeled impairments, disability, or handicap sections.[16] When the Alpha draft of the then-titled ICIDH-2 was compiled, all revision participants had responsibility for all aspects of the entire draft.

In the development of its first revision efforts, the International MH/ ABCD Task Force reviewed SSA's disability determination and the PRTF to ensure the inclusion of appropriate domains and items pertaining to disability for work. Particular attention was paid to the inclusion of the first three B criteria. The items that constitute the first two B criteria are currently located in the ICF activity component (i.e., activities of daily living [B1] and social functioning [B2]) while a major aspect of the third B criterion (i.e., concentration) is located in the impairment section (i.e., sustained attention). The description of the assessment of severity in the mental disorder listings includes extensive lists of both activities that constitute activities of daily living and social functioning, most of which can be found as individual items in the ICF. Because the ICF provides definitions, other parts of the third B criterion—persistence and pace—can be operationalized in one or more of the items contained under the ICF heading "carrying out daily routine" (ICF code d230).

Other aspects of work—responding to current thought in the field of psychiatric rehabilitation—were introduced under the activities component. These items include seeking, maintaining, and terminating jobs. Interpersonal aspects of work are also among the items in interpersonal relationships, such as "relating to persons in authority" and "relating with subordinates," that can be applied to supervisory relationships. Other codes readily apply to relationships with coworkers. The ICF also contains a chapter on general tasks and demands that has items concerning undertaking single and multiple tasks, working independently and in groups, and handling stress. They are worthy of SSA's attention.

Related Assessment and Research Instruments

Research on three instruments related to ICF developed for assessment of, and research on, functioning and disability is in different stages.

---

[16]Impairments were developed under the auspices of the French Collaborating Center; disabilities/activities under the Dutch; and handicap/participation under the North American Collaborating Center (NACC). NACC is housed in the National Center for Health Statistics, Centers for Disease Control and Prevention.

However, these instruments are worth review and consideration. Each is being tested internationally for its psychometric properties.

ICF Checklist The ICF Checklist is an instrument that provides an overview of a person's functioning and disability in the ICF structure. The current version is designed to be completed by a health care worker using any one or a combination of sources of information: written records; primary respondent; other informants; or direct observation.[17] It is essentially an abbreviated list of the items of major interest in health and disability care from the ICF.

The first section asks about demographic information, including occupation and medical diagnosis. This is followed by Part 1a, which queries on impairments of body functions, and Part 1b for body structure. Part 2 contains a short list of activity and participation domains and items, Part 3 contains environmental factors and is followed by a narrative section designed to give a thumbnail sketch of the contextual factors—both personal and environmental—that might have an impact on the person's functioning. The first appendix contains a brief health information questionnaire and is followed by a second appendix that provides guidance to an examiner when interviewing a respondent for completion of the ICF Checklist.

Use of the ICF Checklist is based on the qualifiers found in the ICF rather than a dichotomous rating inferred by the name.

The ICF Checklist offers SSA a basis for a standard form for provision of medical evidence for claims based on all physical or mental conditions. It supplies a potential format for identifying the Listing under which a claim will be adjudicated, the domains in which the claimant is restricted in functioning, and the type of additional medical evidence that would have to be developed. The ICF Checklist could be used from the point of initiation at the SSA DO. The compatibility of ICF for this purpose, in fact, is apparent in that the first two sections of the ICF Checklist contain information that can be applied to and/or used to modify or replace other SSA forms. Because sufficient additional information is collected to assist in various aspects of the disability determination, the ICF format would prevent applicants from "gaming" toward a favorable disability determination. Furthermore, a modification of the ICF to make it more specifically suited to the SSA disability determination could be developed that would include the salient items in the domains of activities that have work-related items. Such a product might provide the basis for providing concrete information upon which to rate the B criteria at the DDS level of review.

---

[17]A self-evaluation version is under development.

WORLD HEALTH ORGANIZATION DISABLEMENT ASSESSMENT SCHEDULE II (WHO DAS II)

In a separate but related activity in the revision of ICIDH and the creation of the ICF Checklist, the WHO and three institutes of the National Institutes of Health (NIH)[18] joined in a cooperative agreement to develop disability assessment instruments for use in clinical settings and in epidemiological surveys. The WHO Disability Assessment Schedule II was designed for use in assessing disability irrespective of health-related etiology. It is calibrated for use to assess disabilities associated with mental as well as physical conditions.

The WHO DAS II is conceptually based on the ICF and queries six domains:

1. understanding and communicating;
2. getting around;
3. self-care;
4. getting along with people;
5. life activities (work and household activities); and
6. participation in society.

The WHO DAS II has undergone testing of its psychometric properties, notably reliability, validity, and sensitivity to change. It underwent testing in approximately 28 centers in more than 18 geographically and culturally diverse countries. Three versions are the product of this endeavor: the full 36-item instrument; a 12-item screener; and a 5-item short form.

Because the WHO DAS II is designed to be used either in conjunction with a diagnostic assessment instrument (such as the Schedules for Clinical Assessment in Neuropsychiatry or the Composite International Diagnostic Interview) or as a stand-alone instrument, the rating of the domains of disabilities is preceded by two sections. The first section collects demographic and other background information.

Twelve screening questions are located in the second section. Like WHO DAS II, these 12 questions are intended to be equally relevant for people with either mental or physical health conditions. Preliminary studies at WHO have shown that this brief questionnaire can predict nearly 90 percent of the variance found in longer versions of WHO DAS II. Ongoing analysis of the WHO DAS II is being conducted that explores its relationship to disability for work. It might behoove SSA to consider additional

---

[18]The three NIH Institutes are the National Institute of Mental Health, the National Institute on Alcohol Abuse and Alcoholism, and the National Institute on Drug Abuse.

research on this—or other brief screening instrument—to be used at the initial stages of the disability determination process to screen out claims based on slight or not-severe limitations in function related to work and to identify those with extremely severe disabilities. Research and data analytic strategies could identify the threshold for slight or not-severe and for extreme or so-severe limitations that would provide an evidence base for this step in sequential evaluation. It might be worth considering applying this standard even prior to the review of A criteria.

RESEARCH ISSUES

There are many compelling issues deserving of research. The committee's second interim report (Wunderlich and Rice, 1998) is replete with them. The research questions are important, timely, and utilitarian. This review of the SSA's disability determination of claims based on mental impairment, the APA study, and the WHO ICIDH-2 suggests additional useful, scientific avenues of investigation. However, at present there is no overarching strategy for identifying and prioritizing research necessary to improve the disability determination for SSA.

In the mid-1980s, mental impairments were added to the Listings, many of which take into consideration functional consequences of an impairment. The committee understands and supports the need to revise and update the Listings to restore them closer to their original purpose. However, the committee is not aware of any attempt to evaluate the currency and consistency of listings, or at least those groups of conditions that account for a significant proportion of the disability rolls. SSA appears to have made the decision to replace the current Listings with an index without any attempt to first evaluate the Listings and use the findings to update them or to guide in developing a new index. SSA should specify the desired levels of specificity and sensitivity and evaluate the current listings against those standards to serve as a baseline for creating the new index (Wunderlich and Rice, 1998, p. 19).

The APA study suggests that the medical component of sequential evaluation for claims based on mental impairment works sufficiently well that only refinements to SSA forms, identification of a period of review, and improvement of the medical evidence (e.g., development of a standard form for basic medical evidence) are warranted. The ICF with its cultural sensitivity and applicability to diverse populations is suggestive of ways of improving both the medical evidence and the SSA forms. These seem relatively small issues in the big picture—when or if the big picture is clear.

What is the big picture for SSA? What are the aspects of the determination process that work well and which are those that do not? Using

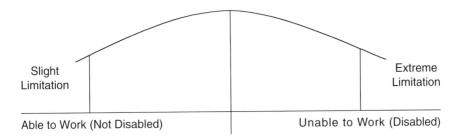

FIGURE 2 Hypothetical distribution of disability for work among SSA claimants.

SSA's existing data, data from the APA study, and findings of the National Academy of Social Insurance, it is possible to identify the major questions raised in the existing five-step sequential process and the proposed four-step sequential process:

- Imagine a bell shaped curve (Figure 2)—a normal distribution of disability for work—among SSA claimants. (It is not truly a normal distribution. SSA data can supply this information.)
- Bisect the curve with a perpendicular line. To the right of the line are those claimants who cannot work, who are disabled for work according to the SSA statutory definition. To the left are those who are not disabled for work.

The first step of claims for both SSDI and SSI benefits based on either mental or physical disorders in sequential evaluation decides whether the claimant is already working. These people are not processed beyond the DO and do not appear above. There are no plans to change the first step.

The second step in the existing sequential evaluation eliminates those with slight limitations (claimants at the far left on the curve). They represent an extreme and are easy to decide on medical evidence. This was not a problematic decision in the APA study. This step is accomplished using the B criteria on the PRTF. No such determination will be made in the proposed SSA revision. This group with slight limitations would be carried along the whole process. The Committee to Review SSA's Disability Decision Process Research in its second interim report recommended that SSA should use a global functional assessment measure to screen out people who do not have severe disabilities (Wunderlich and Rice, 1998, p. 22).

The third step allows benefits to those who are so severely limited (those at the far right of the curve) according to the B criteria of the

Listings as rated on the PRTF. This too is a relatively easy group to identify. It was not considered problematic in the APA study.

One proposed revision to SSA's process would use an Index of Disabling Impairments to make this decision. The committee' second interim report (Wunderlich and Rice, 1998, p. 19) noted that the Index is supposed to be simple enough so that laypersons will be able to understand what is required and to demonstrate a disabling impairment. The Index is being or about to be developed. Yet, there is already a scientifically based finding that the B criteria of mental impairments review work well. Would it not be worth investigating whether those same criteria would work as well for claims based on physical conditions?

Next, disability determination sorts the dichotomous decision of ability or inability to work for the remaining claims in the middle. Without knowing the scientific justification of collapsing steps 4 and 5 of the present sequential evaluation, it seems a reasonable step on face value.

The difficult claims to adjudicate are those close to the line in the middle. This is the group in which false positives and false negatives are most likely to occur. False positives are those who can truly work but are erroneously allowed disability benefits. These individuals will not appeal. Are they likely to stay on the rolls? False negatives are likely to appeal. This is the step in sequential evaluation that the APA study identified as problematic. It is also the step at which SSA has considered using an individualized functional assessment. For claims based on mental disorders, this new assessment would replace the MRFCA.

In the existing sequential evaluation, steps 2 (slight) and 3 (so severe) appear to be quite separate. In fact, they are two decisions made at the same point, using the same information. After claims have had their medical condition identified by the A criteria, the medical evidence is rated using the B criteria. The ratings sort those with slight limitations from those who are very severely limited and those in the middle. This one part of the process actually makes three decisions: denials for slight limitations; allowances for very severe limitations; and those in the middle who need more thought. *If the decisions are correct, it is a very effective and efficient step—the rating of the B criteria.*

What makes sense is to have the DO staff conduct step 1 (is the person involved in substantial gainful activity?) and document the health condition or combination of health conditions that are causing the purported disability for work, and record the onset of the health condition and the onset of the period of disability, thus identifying the period of review. The SSA's form for adults, SSA-3368-BK, contains the applicant's report of these dates and is collected by the DO. Medical evidence can substantiate the health condition and its onset. This would eliminate a time-consuming review with A criteria.

Once medical evidence has been added to the case folder, it might make sense to turn to a brief functional assessment such as the WHO DAS II 12-item screening questions to eliminate the slightly and very severely limited claims as is currently done for claims based on mental disorders. This would settle the claims at both extremes of the distribution. Research here could investigate the applicability of the three B criteria to claims based on physical health conditions and foster a more robust set of B criteria and scale points

For the remaining claims, SSA could build on the APA study and look for the factors that predict difficult-to-adjudicate claims: coexisting mental and physical conditions; onset of less than 12 months; and female claimants. SSA can identify other characteristics from among the claims whose decisions are reversed upon appeal. Additionally, scores or patterns of ratings on the WHO DAS II 12-item screening questions might distinguish routine from difficult-to-determine claims. These will be the claims likely to be closest to the border of able or not able to work. The APA suggested that these difficult claims might be best served by a review by a panel of Reviewing Medical Consultants.

Information SSA Already Has

The answers to the following questions reside either in the various datasets of SSA or in reports submitted to it by such informed sources as the National Academy of Social Insurance. These data can identify the strengths and weaknesses in the existing process.

1. SSA can identify the distribution of disability among its claimants. It should do this for three groups: all claims; claims based on physical conditions; and claims based on mental conditions. SSA may suggest reasons why this should be done for both disability programs and SSDI and SSI separately for the three groups of claimants. The reason for conducting separate analyses for physical and mental conditions is to understand the generalizability of the APA findings to claims based on physical disorder. SSA may also consider separate analyses of other problematic categories of physical disorders.

2. SSA can identify the magnitude of importance for slight and very severe disability decisions. How many of these decisions are made? What proportion of claims do they represent? How many denials based on slight limitation are appealed and reversed? How many very severely limited people leave the rolls and return to work? How long does a claim take to assess when it is denied at step 2? How long does a claim take to adjudicate that is allowed at step 3?

3. What are the characteristics of claims that take a long time to adjudicate?
4. What are the characteristics of claims that are denied at steps 4 and 5 that are appealed and have the disability decision reversed?
5. Can SSA identify the types of medical evidence that are most difficult and/or take the most time to obtain?

The above information can begin to establish priorities for additional research. The information tells us what is working well and what is problematic. *If it is working, why change?*[19]

Nonetheless, with improvements in the submission of medical evidence, the number and percent of claims that can receive a disability determination at steps 2 and 3 are likely to be increased, thus reducing the number that continue in the determination process. Information based on ICF items might be requested as part of a standard submission of medical evidence.

The next issue is to sort the claims remaining after the slightly limited and very severely limited have been handled. For these claims, additional functional assessment—but no consideration of age, education, and work history—is planned in the proposed process. This is where new functional assessment forms or instruments are required. Research would have to explore these alternatives and the medical evidence needed for the assessment.

The above statements are consistent with Recommendation 4-1 made by the IOM committee in its second interim report (Wunderlich and Rice, 1998, p. 21).[20]

Problems Already Identified by Research

The lack of sufficient medical evidence and the low quality of the medical evidence that is provided were identified as serious issues by participating psychiatrists in the APA study. This was their impression even though the claims had been through an SSA review for the quality of the evidence. In other words, they may have been better or more complete than average claims adjudicated in the DDSs. In addition to the APA's

---

[19]This may be the information to which the committee alluded in its second interim report (Wunderlich and Rice, 1998, p. 13) in discussing the nature and extent of the problem with the disability decision process. The above information may provide sufficient information to act as a needs assessment.

[20]The committee recommended that early in its redesign effort, the SSA should specify how it will define, measure, and assess the criteria it will use to evaluate the current disability determination process, as well as any alternative processes being developed.

recommendation that a standard form for medical evidence be designed and used nationally, other medical evidence collection procedures might be reviewed. For example, although SSA is seeking to streamline its process, something may be gained by collecting medical evidence in two stages: the first stage for slight and very severe decisions based on a modification to the first three B criteria, the second for more in-depth assessment of the remaining claims, notably those that are expected to be inherently difficult, near the border of able or unable to work. What is needed? Which items in ICF can guide the development of medical evidence forms(s)? Have some states developed standard medical evidence forms, and if so, how are they working?

The APA study also identified problems in reviewing claims based on mental disorder when a physical condition was present. This issue might be addressed by a de novo consideration of the confluence of review of claims based on mental and physical disorders—*the basic requirements of work for the two groups do not differ*. Can a useful set of work-related factors be identified across disorder groups? For example, the items in ICF include general tasks and demands that are the basis for work-related activities (ICF d210 through d299); communications (ICF d310 through d349); interpersonal interactions and relationships (ICF d710 through d7109, d740 through d7409); and tasks needed to find, get, and keep a job (ICF d845 through d8459). A review of the ICF along with SSA's regulations and relevant Program Operations Manual System might reveal additional items. Any rating of these items would be consciously made in view of the claimant's health condition and other criteria outlined in the SSA definition of disability.

How might the first three B criteria for claims based on mental disorders and their rating be approved and made applicable to people claiming disability based on physical disorders? Again, the ICF offers a rich resource both for the items and for rating the items. The fourth B criterion might be revisited along with the C criteria to create a method to acknowledge that certain individuals appear not to be disabled for work but are only compensated in their existing environment or accommodated to their current level of stress. A change to a work environment would destroy their fragile state.

Of note is that there are no analogous B criteria for the Listings for physical conditions. In general, functional restrictions in the Listings for physical conditions are limitations in the function of a body part, such as motion of a joint. Only in the Residual Physical Functional Capacity Assessment (analogous to the MRFCA) are there any physical functioning ratings that apply to all of the physical categories. Not all of them are limitations in *activities* such as lifting and carrying; some are limitations of specific *body parts or systems*, such as visual acuity. Many of the physical limitations are redundant with functions of body parts of the Listings of physical impairments (e.g., reaching in all directions [Residual Physical Functional Capacity Assessment] and range of motion in joints [physical Listings]).

STRATEGY FOR DEVELOPING AN SSA DISABILITY RESEARCH PROGRAM

Prior to developing a program of research, identifying issues for research, and specifying research mechanisms and methodologies, a strategy for establishing priorities needs to be created. At present, compelling research questions are vying for attention. The IOM committee may wish to consider establishing a working group to draft an outline of work that addresses the following three issues preliminary to creating priorities for research:

1. Using SSA information, identify *weak* points in the existing *adjudication process* (e.g., slight limitation denials) and identify *strong* points that can be built on or generalized across claims based on physical and mental disorders:

   • Develop a plan for "fixing" the sequential process. For example, if the first two steps work well, keep them. Then go to the third step to look at problem areas (keeping in mind that for claims based on mental disorders, steps 2 and 3 are not readily distinguishable as steps only for the decision that is made).
   • Acknowledging that there will always be *inherently difficult claims*, develop a way of preselecting them and creating a process for handling them (e.g., panel of reviewers).

2. Explore the differences between the *adjudication materials* for claims based on mental disorders and those used on physical disorders to see if the *strengths* identified by the APA study might not apply to the adjudication of physically disordered claimants, and attend to the *weaknesses* as well:

   • If the A criteria are to be eliminated as has been suggested at times, identify some method of documenting the association of the health condition with the disabled state.
   • Explore the factors that need to be included in the medical evidence and how the medical evidence will interface with SSA assessment forms.

3. Review the *ICF, ICF Checklist, WHO DAS II, and WHO DAS II screening questions* to identify which components can: be used to create a standard form for the submission of medical evidence; can be used as a screen of slight and very severe claims; and can provide the basis for a fuller, more detailed version of the functional limitation criteria for both physical and mental conditions, and conduct a thorough investigation of the scale points and ratings as they calibrate to disability decisions (not severe, so severe, and disabled for work) for both physical and mental conditions.

The product of this working group has the potential to lead to a cogent research agenda that allows the inclusion of additional important research topics. It will also make evident the type of research that needs to be conducted, such as clinical trials or instrument development research, and the methods and mechanisms by which this research should be conducted. It also might suggest partners in this important national and federal endeavor.

RECOMMENDATIONS

There are four overarching areas in which recommendations can be grouped: (1) disability determination; (2) identifying priorities; (3) conceptual, taxonomic, and assessment resources; and (4) research agenda. All of the following recommendations are empirically based and made with the proviso that each be accompanied by research in an iterative process to provide a scientific base to identify and substantiate improvements. These recommendations are based on the disability determination of claims based on mental disorders and on research of the process, standards, and guidelines used in the adjudication of claims based on mental disorder.

Disability Determination

1. Existing research on disability claims-based mental disorders finds that **no *major* change** to sequential evaluation or to the standards and guidelines used in this process (notably the Listings of Mental Impairments and the PRTF) is warranted—only refinements.
2. Because basic *work requirements* are **independent** of *health conditions*, the adjudication of functional capacity for work should be the same for claims for disability benefits irrespective of the type of health condition (i.e., physical or mental disorder).
3. The current process of sequential evaluation should be maintained with the following modifications:

- There should be four steps to sequential evaluation: the current steps 4 and 5 should be combined.
- The first judgment in the medical review should be identification of the date of onset and period of review.
- As long as the health condition causing the disability for work (i.e., the A criteria of the Listings of Mental Impairment) can be identified, substantiated, and linked with functional limitations, the focus of these criteria can be shifted to a more thorough and substantive evaluation of functional capacity for work.
- Based on an evaluation of functional capacity for work, sequential evaluation should screen out claims that are not severe and those that are so severe. These two disability decisions (i.e., denials for not severe and allowances for so severe) are currently steps 2 and 3 in the sequential evaluation. These two medical judgments of disability status are focused on the two extremes of the distribution of disability claims.
- For the remaining claims, SSA should apply clinical and demographic factors that identify difficult-to-adjudicate claims and select these claims for review using a panel process that combines additional functional assessment and nonmedical factors as the last step in sequential evaluation.
- For claims not identified as difficult to adjudicate, the final step in sequential evaluation should be applied, which combines additional functional assessment and nonmedical factors.

Identifying Priorities

1. SSA should analyze its existing data to identify and prioritize areas for revision and refinement.

- SSA should compare the magnitude of its caseload and the decisions made at each step in sequential evaluation for all claims for disability benefits in the SSDI and SSI programs, all claims for disability benefits based on physical conditions in the SSDI and SSI programs, and all claims for disability benefits based on mental conditions in the SSDI and SSI programs.
- SSA should identify and prioritize the weak points and strengths in the existing sequential evaluation for claims based on physical conditions and for those based on mental conditions by examining the magnitude of decisions made at each step in the process and by the proportion of denied claims that are successfully appealed. Steps 2, 4, and 5 can be evaluated in this fashion.

- SSA can identify claimants awarded disability benefits at step 3 who leave the rolls and return to work.

2. Using existing data, SSA can identify the characteristics of claims that are difficult to adjudicate.

- Based on the APA study findings, SSA can explore clinical and demographic factors that predict difficulty in reaching a disability determination.
- SSA can identify inherently difficult claims by specifying the characteristics of those that take a long time to reach a disability determination.
- SSA can examine the types of claims that are likely to be reversed on appeal to identify inherently difficult claims.

Conceptual, Taxonomic, and Assessment Resources

- The WHO's ICF, an international classification of disablement and functioning, provides a culturally sensitive, research-based, rigorous yet flexible conceptual foundation for revisions to SSA's disability determination for claims based on mental and physical conditions. One of its specified applications is for Social Security disability benefit programs.
- The ICF conceptual model is consistent with the definition of disability in the Social Security Act and with the process that puts disability determination into operation.
- Domains and items in ICF are readily utilized in modifications to existing SSA disability determination forms in sequential evaluation. They would enrich SSA's conceptual development and assessment of functional capacity to work.
- The ICF has related disablement assessment and research instruments that can be modified for use as standard forms for medical evidence and in sequential evaluation:

1. The ICF Checklist can be used as the basis for the development of a universal form for submission of medical evidence. It should be reviewed for possible inclusion of additional items related to work activities.
2. The WHO DAS II contains six domains, five of which relate to work as currently conceptualized in the Listings/PRTF B criteria applied to claims based on mental conditions. This instrument could be supplemented with additional physically exertional items and used as the residual functional capacity assessment

for both mental and physical conditions in the last step of sequential evaluation.

3. Additional research should be conducted on the WHO DAS II as adapted above, and on its scoring and scale point definitions, to provide a scientific judgment about work capacity. (At present, the part of the MRFCA form that is used in the determination is a narrative summary. The 20-item ratings are considered a worksheet.)

4. The WHO DAS II 12-item screening questions is composed of two items from each of the six domains. The scaling of these items can be tested against the current B criteria thresholds for identification of the two extremes not-severe and so-severe claims. They can then be used for claims based on physical and mental conditions to make disability determinations at steps 2 and 3 of sequential evaluation.

Research Agenda

1. A working group of researchers from SSA, the National Institute of Mental Health, other federal agencies, and knowledgeable researchers from the private sector should be organized to develop a research agenda to

- review the ICF, ICF Checklist, WHO DAS II, and WHO DAS II 12-item screening questions for needed modifications for applicability in SSA sequential evaluation;
- refine assessment instruments for the evaluation of disability for work based on physical and mental conditions for use in sequential evaluation;
- calibrate scaling of assessment to disability for work based on physical and mental conditions;
- create a standard medical evidence form;
- review SSA databases to analyze existing data and identify priorities for research; and
- review IOM recommendations for research.

2. All refinements to the existing system should be based on research findings.

3. The development of revisions should be an iterative process with research findings providing an empirical base.

4. Appropriate research methodologies should be identified for the diverse research issues.

In summary, the evaluation of claims for disability benefits is a complex and difficult task. This task is also a small component of the many extrinsic factors that have bearing on the shape of the disability program, such as long- and short-range economic factors, the changing characteristics of the general population and labor force, and the general priority and ideology held regarding people with disabilities. This paper has limited its focus to the medical review of claims for disability benefits based on mental conditions and has been informed by the APA's evaluation of SSA's standards and guidelines used in disability determination. With the recent revision of WHO's ICF and the development of related disability assessment and research instruments, a new and valuable resource has become available for use in modifications of the tools used in sequential evaluation.

Building on the research database provided by the APA study, the development of the ICF as a conceptual model, and on a classification system of disablement and functioning and its related instruments, it is possible to suggest recommendations for a research-based agenda to refine sequential evaluation and the standards and guidelines that implement disability determination for claims based on both mental and physical conditions. This is made possible because basic work requirements are consistent across disorder types. Each of the recommendations that have been made is intended to be a component of an explicit research plan developed by an interagency working group.

Finally, it is important to note that all recommendations are made within the context of the SSA's definition of disability. There are no indications that any change to the statutory definition should be considered.

REFERENCES

Americans with Disabilities Act. 1990. Public Law 101–336.

American Psychiatric Association. 1980. *Diagnostic and Statistical Manual of Mental Disorders* (3rd Ed.) Washington, DC: American Psychiatric Press.

Pincus HA, Kennedy C, Simmens SJ, Goldman HH, Sirovatka P, Sharfstein S. 1991. Determining disability due to mental impairment: APA's evaluation of Social Security Administration guidelines. *American Journal of Psychiatry* 148(8):1037–1043.

Pope AM, Tarlov AR. 1991. *Disability in America: Toward a National Agenda for Prevention.* Washington, DC: National Academy Press.

Social Security Act. as in effect January 2001.

Social Security Administration. 2001. *Disability Evaluation under Social Security. Listing of Impairments Adult* [Online]. Available: http://www.ssa.gov/disability/professionals/bluebook/AdultListings.htm [accessed September 24, 2001].

Social Security Administration Regulations. 20 CFR Ch.III 404.1520 [Online]. Available: http://www.ssa.gov/OP_Home/cfr20/404/404-0000.htm [accessed September 24, 2001].

Social Security Administration Regulations. *20 CFR Ch.III 404.1520a* [Online]. Available: http://www.ssa.gov/OP_Home/cfr20/404/404-0000.htm [accessed September 24, 2001].

World Health Organization (WHO). *ICF Checklist* [Online]. Available: http://www3.who.int/ icf/icftemplate.cfm [accessed September 24, 2001].

WHO. 1980. *International Classification of Impairments, Disabilities, and Handicaps: A Manual of Classification Relating to the Consequences of Disease*. World Health Organization, Geneva.

WHO. 1993. *International Statistical Classification of Diseases and Related Health Problems: Tenth Revision*. World Health Organization, Geneva.

WHO. 2001. *International Classification of Functioning, Disability and Health*. World Health Organization, Geneva.

WHO [Online]. Available: http://www.who.int/icidh/whodas/index.html [accessed September 24, 2001].

Wunderlich GS, Kalsbeek WD, eds. 1997. *Disability Evaluation Study Design: First Interim Report*. Washington, DC: National Academy Press.

Wunderlich GS, Rice DP, eds. 1998. *The Social Security Administration's Disability Decision Process: A Framework for Research. Second Interim Report*. Washington, DC: National Academy Press.

Survey Design Options for the Measurement of Persons with Work Disabilities[1]

Nancy A. Mathiowetz, Ph.D.[2]

INTRODUCTION

As noted in a recent publication, "Capturing the essential medical, physical and social aspects of disability by means of survey data is a difficult task" (Altman, 2001). The complexity stems, in part, from differences in conceptual models of the enablement–disablement process and alternative interpretations of various conceptual models. The incongruity between various conceptual models of disability and the Social Security Administration's (SSA's) model based on its statutory definition of work disability adds further complexity to the measurement process when one is particularly interested in estimating the pool of potential applicants or the number of those who would be classified as persons with work disabilities as a result of SSA's benefits decision process. The former requires that the survey questions represent an accurate operationalization of the SSA statutory definition; the later requires additional information related to the SSA decision process.

Given the complexity of the phenomena of interest, agreement with respect to the conceptual model does not imply consistency in the

[1]This paper was originally prepared for the committee workshop titled "Workshop on Survey Measurement of Work Disability: Challenges for Survey Design and Method" held on May 27–28, 1999, in Washington, D.C. (IOM, 2000).
[2]Nancy Mathiowetz is Associate Professor, Joint Program in Survey Methodology, University of Maryland.

operationalization of the concept. Alternative operationalizations are evident in the variety of types and numbers of questions used in various surveys to measure functioning and participation. Empirical evidence suggests that even minor variations in how one operationalizes the concept of disability can result in significant variations in the estimates of the population of persons with disabilities in the United States (see, for example, McNeil, 1993). Consideration of alternative design options for household-based survey measures of the population of persons with work disabilities requires that SSA invest in understanding how divergent measurement may affect estimates of the potential pool of applicants for SSA benefits as well as estimates of the population of potential beneficiaries.

The task of using household-based surveys for the measurement of persons with work disabilities is further complicated by the dynamic state of the field at the time of this writing. A number of research activities, both within the United States and internationally, related to the measurement of disability via the survey interview process, will most likely result in major changes to question wording and questionnaire design over the next decade. Several federal statistical agencies, including, but not limited to, the Bureau of Labor Statistics and the Bureau of Justice Statistics, are testing questionnaires to meet legislative or executive branch mandates related to the production of statistics by disability status. The adoption of the second revision of the International Classification of Impairments, Disabilities, and Handicaps (renamed ICIDH-2: International Classification of Functioning, Disability, and Health) by the 54th World Health Assembly provides a classification system and framework for the development of disability measures in surveys. Much of the international research on disability measurement is focused on the development of valid and reliable instruments that map conceptually to the ICIDH-2, including measurement of the environment. Other research activities are attempting to address gaps in the current state of knowledge concerning the measurement error properties of disability statistics

In light of the challenges facing SSA, this paper attempts to outline and discuss design options related to the measurement of persons with work disabilities and the measurement of persons eligible for SSA benefits. The paper examines a broad range of alternatives for SSA to consider, ranging from the development of its own data collection and measurement system[3] to the use of other federal data collection efforts for ongoing monitoring. We begin by examining the disparities between the definition of persons with work disabilities used by SSA and the models

---

[3]The use of the term "measurement system" is intended to refer to an integrated, long-term data collection system as compared to periodic, uncoordinated data collection efforts.

underlying the measurement of disability in other surveys. In addition, we review what is currently known with respect to the error properties of survey-based estimates of the population of persons with work disabilities.

CONCEPTUAL ISSUES TO CONSIDER IN THE MEASUREMENT OF PERSONS WITH WORK DISABILITIES

As noted above, one of the issues of concern in the measurement of persons with work disabilities involves differences in the conceptual models underlying the measurement. The Social Security Act defines disability as the "inability to engage in any substantial gainful activity by reason of a medically determinable physical or mental impairment which can be expected to result in death or can be expected to last for a continuous period of not less than 12 months" (Section 223(d)(1)). The SSA definition implies a direct relationship between an individual's attributes—specifically related to pathology, impairment, and functional limitation—and work disability. In contrast, Nagi (1991, p. 317) and other contemporary theorists characterize disability as a ". . . relational concept; its indicators include individuals' capacities and limitation, in relation to role and task expectation, and the environmental conditions within which they are to be performed." As noted by Jette and Badley (2000, p. 17):

> The fundamental conceptual issue of concern is that a health-related restriction in work participation may not be solely or even primarily related to the health condition itself or its severity. In other words, although the presence of a health condition is a prerequisite, "work disability" may be caused by factors external to the health condition's impact on the structure and functioning of a person's body or the person's accomplishments of a range of activities.

Most measures of work disability currently in use in U.S. federal surveys assume (or imply) that work disability relates to an individual's attributes with respect to functional limitations; almost all such questions leave it to the respondent to attribute his or her labor force participation to an underlying health condition. However, the movement in the measurement of persons with disabilities and persons with work disabilities is toward measures that incorporate an understanding and assessment of the external factors that influence participation by individuals in work. It is conceivable that as new measures for the assessment of persons with work disabilities are developed and adopted in ongoing federal data collection efforts—that is, measures that incorporate an assessment of the environment (including accommodations, adaptations, and barriers)—the discrepancy between survey-based estimates of the population with work disabilities and the population eligible for SSA benefits will increase. To the extent that SSA may rely on survey-based measures of disability drawn from non-SSA-sponsored surveys, it will be imperative for SSA to understand how such measures relate (both conceptually and statistically) to the statutory definition of work disability.

METHODOLOGICAL ISSUES IN THE MEASUREMENT OF WORK DISABILITY IN SURVEYS

As was evident in the report of the Workshop on Survey Measurement of Work Disability (Mathiowetz and Wunderlich, 2000), the measurement of persons with work disabilities via household-based surveys is subject to various sources of error that may or may not result in bias in the estimate of interest. Although not unique to the measurement of persons with disabilities, the complexity of the concept as well as the very nature of the phenomena of interest suggests a need to be particularly vigilant with respect to the potential impact of errors of both observation and nonobservation on estimates of the population. There is little to no research examining the impact of nonobservation (both noncoverage and nonresponse) on estimates of the population of persons with disabilities or persons with work disabilities. One could speculate on the non-ignorable nature of nonresponse of such estimates, hypothesizing that persons with disabilities are less likely to participate in household-based surveys. For example, the nature of the disability (e.g., sensory) may limit participation for particular modes of data collection. Empirical investigations are needed to understand the extent to which errors of nonobservation bias survey-based estimates of persons with disabilities.

Similarly, we have some empirical data to indicate that survey-based estimates of persons with disabilities may be plagued by problems of measurement error. The empirical evidence suggests that factors as diverse as the mode of data collection, the sponsorship of the survey, the nature of the respondent (i.e., whether the individual reports for him- or herself or is reported for by someone else in the household), the specific question wording, and questionnaire context, as well as the order of the questions, may affect the estimates of the population and the stability of those estimates (see Mathiowetz, 2000). However, to date, little empirical research has isolated the effects of specific design features. For example, questions administered as part of the National Survey of Health and Activities (NSHA) may yield very different estimates of the population when administered using a different mode of data collection or when administered as part of a different study (which may reflect a change in context, question order, and sponsorship).

This lack of research with respect to the impact of either errors of nonobservation or errors of observation is critical in thinking about alternative design features for the measurement of persons with work disabilities. For example, let us assume that SSA decides to conduct a NSHA-type survey every k years, with monitoring of the pool of eligible applicants based on a subset of the NSHA questions in the form of a topical module administered as part of the Survey of Income and Program Participation (SIPP) in the intervening years. Most likely, the two surveys—that is the NSHA-type survey and the SIPP—will vary on a number of essential survey design features, including survey context, mode of data collection, and survey sponsorship, all of which could impact the level of reported work disabilities. The two surveys may also differ with respect to coverage of the U.S. population. Differences in the rate of nonresponse, or even the mix of the types of nonresponse (i.e., refusals versus noncontact), could lead to differential nonresponse error across the two surveys. Without a systematic program of research that addresses the relative effects of differential nonresponse as well as the effects of various design features on levels of measurement error, SSA will be without empirical-based information with which to determine whether year-to-year variations in estimates are attributable to true change or differences in the design of the two studies. Considerations of alternative survey designs for the measurement of persons with disabilities cannot ignore the potential impact of either measurement error or errors of observation on estimates.

ONGOING DEVELOPMENTS IN DISABILITY MEASUREMENT

Because notions of disability and models of influences on disability are constantly changing, any ongoing system to monitor the phenomena must be able to adapt and change over time. This can be accomplished only with ongoing monitoring of the scientific endeavors in the field and investment in new methods of measurement. Much of the present research in the measurement of disability in surveys is focused on developing question items that map conceptually to the International Classification of Functioning (ICF, formerly ICIDH and ICIDH-2) (World Health Organization, 2001).

The ICF model advocates the measurement of disability on a continuum as opposed to the binary categories of disabled and nondisabled that have predominated in the survey measurement of disability. In addition, the model depicts disability as an interaction between a person, his or her health condition, and the environment in which the person lives, an integration of medical and social models of disability (biopsychosocial model). The nature of the physical and social environment can either limit or assist so as to result in various levels of activity and participation by an individual. As such, the framework differs from theoretical models that depict disability as a process beginning with impairment and ending with social role or behavioral restrictions or models that focus on disability as merely a functional limitation, that is, the restriction in physical functional activity and task activity associated with the impairment (Altman, 2001). In addition, the use of neutral terminology (as opposed to negative terminology such as handicap or disability) is emphasized in the ICF framework.

The ICF focuses on nine domains: (1) learning and applying knowledge, (2) general tasks and demands, (3) communication, (4) mobility, (5) self-care, (6) domestic life, (7) interpersonal interactions and relations, (8) major life areas, and (9) community, social, and civic life. Within these nine domains, body function or capacity independent of environment as well as performance (which is dependent on environment) are to be measured. For example, an individual may have a latent allergy (body function) that manifests itself only when the person is exposed to the allergy agent, which may or may not therefore affect performance. Performance includes both execution of actions by an individual (activity) and involvement in life situations such as work (participation).

The release of the ICF in the spring of 2001 has resulted in a number of research activities related to the design of questionnaires that can be mapped to the ICF framework. Much of this research focuses on question wording to measure activity (and the use of assistance in the performance of activities) and participation (both extent of participation and satisfaction with participation). There is a great deal of research interest related to the development of a single reliable and valid question that could be proffered for use in censuses internationally. In addition, questionnaire design research has focused on the construction of both short- and long-form questionnaires with known measurement error properties. The movement from dichotomous response options to continuous response classification has led to questions as to the impact of cutpoint decisions on estimates of the "disabled" population as well as their impact on the distribution of the characteristics of the population.

SOCIAL SECURITY ADMINISTRATION'S DISABILITY SURVEYS: HISTORICAL PERSPECTIVE

A key component of good fiscal management for the Social Security Administration is having sufficient information to understand and predict growth in the pool of persons eligible for disability benefits as well as understand the factors that impact the application process, including motivation to apply for benefits. Medical models of disability have historically been insufficient to explain unexpected growth in the size of the applicant pool. Factors extrinsic to the benefits programs, for example, cyclical changes in the economy, as well as social and cultural issues, have in the past resulted in changes in acceptance rates and unexpected increases in program expenditures. One means by which to understand the magnitude and characteristics of the pool of eligibles, as well as the intrinsic and extrinsic factors that lead to applying for benefits, is to develop an ongoing surveillance system related to work disability.

The idea of a measurement system related to understanding the incidence and prevalence of those eligible for benefits and those applying for benefits is not a new idea. Between 1966 and 1978, the SSA sponsored a number of data collection efforts designed to measure the prevalence of persons with work disabilities. The first of these surveys was the 1966 Social Security Survey of Disabled Adults. The survey consisted of an area probability sample drawn from seven different frames, including a frame of Social Security disability beneficiaries, denied applications for disability benefits, and disabled recipients of public assistance (Haber, 1973). The survey consisted of a two-stage design; the first stage was a screening interview of 30,000 households to identify adults ages 18 to 64 years of age with limitations in their ability to work. Personal interviews were then conducted with those identified in the screening, approximately 8,300 individuals. Individuals were classified according to the extent of their self-reported capacity for work: severely disabled (unable to work altogether or unable to work regularly), occupationally disabled (able to work regularly, but unable to do the same work as before the onset of disability or unable to work full time); and secondary work limitation (able to work full time, regularly, and at the same work, but with limitations on the kind or amount of work performed). The study provided information on the prevalence of persons with work disabilities among those ages 18 to 64, as well as information about access to and utilization of health services, income, and demographic characteristics.

The second major survey of disability was the 1972 Survey of Disabled and Nondisabled Adults. The survey included both disabled and nondisabled adults of working age and focused on revisions of the estimates of prevalence as well as factors associated with the development and duration of disability (Bye and Schechter, 1982). Those interviewed in 1972 were reinterviewed two years later to examine changes in both disability status and economic status, as well as the relationship between changes in disability and economic status and entitlement under both the Social Security Disability Insurance (SSDI) and the Supplemental Security Income (SSI) programs.

The large growth in the number of SSDI beneficiaries between 1966 and 1975 was the impetus for the 1978 Survey of Disability and Work. As noted in the documentation for the survey (Bye and Schechter, 1982, p. 2):

The rate of growth of the DI (disability insurance) program gives rise to the question of what accounts for the increase. Investigations of changes in labor market conditions and changes in the application of program eligibility criteria address this problem at a macrolevel. This survey allows for a complementary investigation of the reasons for growth with the individual as the unit of analysis. The policy focus of the survey is on the decision to apply for disability benefits.

Between 1969 and 1979, the SSA sponsored the Retirement History Survey, a longitudinal survey designed to understand the conditions under which persons decided to take Social Security benefits before reaching age 65. Although not a disability survey per se, detailed information concerning health and work limitations was collected during the six waves of data collection, and early analysis of the data indicated the importance of health problems as precursors of early retirement (Bixby, 1976).

In addition to surveys sponsored by the Social Security Administration, scholars of the disability process and the disability application process have relied on longitudinal data collection efforts such as the National Longitudinal Study (NLS) and the Health and Retirement Survey (HRS) to understand factors associated with early withdrawal from the labor force. As noted by Sheppard (1977, pp. 163–164):

> The NLS project, and the type of analysis it makes possible, has a value not associated with the usual cross-sectional project in that it provides an opportunity to make *predictions* regarding *subsequent* work or life status. It is also important to make the point that, despite the criticisms that have been made regarding the utility of self-reported health status the individual's own judgment of his or her health status or work capacity at one point in time is a useful and reliable predictor of subsequent labor force or life status (emphasis in original).

Despite the richness of the data resulting from these various survey efforts, the survey efforts do not permit the analyst to understand fully both the individual factors and the environmental factors that result in a person's shifting from the status of potential applicant to actual applicant. Although previous research has permitted examining macrolevel relationships between economic changes and changes in the size of the applicant pool (e.g., Yelin et al., 1980; Bound and Waidmann, 2000), understanding the contributions of both individual and environmental characteristics to microlevel decisions to apply requires data that are longitudinal and capture information related to an individual's decision process.

ONGOING MEASUREMENT OF PERSONS WITH WORK DISABILITIES: THE DEVELOPMENT OF A WORK DISABILITY MEASUREMENT SYSTEM

Developing a disability measurement system is not dissimilar to the design of a complex survey consisting of multiple components (e.g., the National Health and Nutrition Examination Survey [NHANES], which includes a household-based survey and a medical examination coupled with laboratory testing); for each, one should begin with a clear statement of the objectives of the system and a description of the measures of interest. Once the objectives are established, system designers can focus on the necessary components and operation of the system to meet those objectives (e.g., data sources and frequency of collection). Questions concerning the population to be studied, the frequency and period of data collection, the information to be collected, determination of the provider or providers of information (e.g., household respondents, abstracts from administrative records), and decisions concerning analysis, frequency of reporting, and dissemination of information should be addressed in the design of the system .

The utility of a measurement system is a function of the extent to which the data are used to make decisions, set policy, or implement changes. An assessment of a system's utility should be evaluated in light of the objectives of that system; for example, to what extent does the system permit the detection of changes in the rate of application for disability benefits?

In addition to assessing the utility of a measurement system, other attributes of a well-designed system include its simplicity (in both structure and operation), flexibility, and sensitivity and specificity (to accurately detect cases and to distinguish true positives from false positives); the representativeness of the population being studied; and the predictive value of the system.

A critical element of a measurement system is to clearly define and identify a "case." The disability definition for entitlement of benefits is the same for both the Title II Disability Insurance Program and the Title XVI Supplemental Security Income program, although other requirements differ. As noted earlier, disability is defined under the two programs as "inability to engage in substantial gainful activity because of any medically determinable physical or mental impairment lasting at least 12 months." Of interest in a disability surveillance system is not simply measurement of prevalence and the socioeconomic conditions linked to disability, but also understanding both the individual and the environmental factors that lead to changes in the SSA benefits application process. The system will need measures of the prevalence of the eligible pool as well as measures that predict application. Key to such a system will be sufficient data to understand macro- and microlevel factors that distinguish participating and nonparticipating eligibles. A work disability measurement system will have to include means for modeling the decision process from both the demand side (the individual) and the supply side (the Social Security Administration).

DESIGN OF A WORK DISABILITY MEASUREMENT SYSTEM

The design of a work disability measurement system must consider the analytic needs of the system and the impact of alternative design options on meeting those analytic goals as well as the impact of various sources of survey error (should the design include the use of household- or provider-based surveys). Among the design issues the system will have to address are the following:

- data source or sources;
- cross section versus longitudinal design;
- periodicity;
- mode of data collection;
- self versus proxy response status; and
- specific wording of question, response option presentation, and overall context.

Data Source or Sources Among the various data sources that could be included, alone or in combination, in the design of a disability surveillance system are household-based survey data, provider-based survey data, administrative record data, and physical examination data. Among the options with respect to household- or provider-based survey data are stand-alone surveys that permit rich and deep national data on the size of the disabled population (e.g., similar to the NSHA, which is currently being sponsored by the SSA), survey modules administered as part of some preexisting data collection effort (e.g., a supplement to the Current Population Survey or the SIPP), or the incorporation of a limited number of questions on existing national surveys, for example, the National Health Interview Survey (NHIS) or the Behavioral Risk Factor Surveillance System. Each of these options with respect to household- or provider-based surveys has implications for the error properties of the resulting estimates, including coverage, sampling, nonresponse, and measurement error. In addition, consideration much be given to the costs associated with obtaining data from alternative sources. The use of administrative record data potentially suffers from similar sources of error, including operational definitions of disability that are incongruent with those of the Social Security Administration.

Cross-Sectional Versus Longitudinal Design As noted above, a longitudinal design permits researchers to address analytic capabilities that are not possible with repeated cross-sectional designs, especially those related to the decision to apply for benefits, including both individual factors that influence the decision and the impact of environmental and macrolevel changes (e.g., economic) on the decision to apply for benefits. Longitudinal designs require that additional decisions be made concerning the length of the panel (i.e., the number of years individuals are followed), the frequency of data collection, and the means for following individuals who move.

Use of a panel survey design, with repeated measurements with the same individuals, facilitates more efficient estimation of change over time (compared to the use of multiple cross-sectional samples). However, panel designs may be subject to higher rates of nonresponse (cumulated across waves of the data collection) or panel conditioning bias, an effect in which respondents alter their reporting behavior as a result of exposure to a set of questions during an earlier interview.

Periodicity If survey data are collected, how often should data collection occur? What are the ramifications of more frequent or less frequent data collection for the utility of the data? How is periodicity affected if one decides to utilize a longitudinal design? For repeated cross-sectional data collection?

Mode of Data Collection For survey data collection, a decision as to the mode or modes of data collection will have to be addressed. Little is known about the effect of mode of data collection on the measurement error properties of self-reports of disability and impairments. Selection of mode or modes of data collection involve a complex decision concerning costs, response rate objectives, and measurement error. With respect to costs, face-to face-data collection is significantly higher than other modes. It is for this reason that several federal surveys involving panel designs of households have moved toward mixed modes, with the initial interview conducted face to face and subsequent interviews conducted either by telephone or face to face. Face-to-face data collection is often considered the preeminent mode for data collection, due in part to the opportunity to gather data that are not feasible via other modes (e.g., physical measurements, interviewer observation) and in part to the perception that face-to-face data collection continues to achieve higher rates of response than other modes. However, one must consider that mode comparisons of response rates are confounded by survey sponsorship, with the federal government in the United States conducting or sponsoring most of the face-to-face data collection. The one consistent finding with respect to the effect of mode of data collection suggests that to the degree the information is considered sensitive or socially undesirable, one is more likely to collect more accurate data via self-administrative modes of data collection. For example, although the National Household Survey of Drug Abuse is conducted as a face-to-face interview, questions concerning illicit drug use and other sensitive behaviors are reported via self-administration. The choice of multiple modes of data collection may be desirable from the perspective of reducing coverage bias (e.g., dual-frame sampling designs) or improving response; however such a design decision to reduce errors of nonobservation may come at the expense of an increase in measurement error.

Self and Proxy Response Status Should only self-response be accepted for household surveys of disability? If so, what are the ramifications on nonresponse bias? If proxy responses are accepted, what impact does this design choice have on the measurement error properties of the reporting of disability?

The use of proxy reporters—that is, asking individuals within a sampled household to provide information about other members of the household—is another design decision that is often framed as a trade-off among costs, sampling errors, and nonsampling errors. The use of proxy informants to collect information about all members of a household can increase the sample size (and hence reduce the sampling error) at a lower marginal data collection cost than increasing the number of households. The use of proxy respondents also facilitates the provision of information for those who would otherwise be lost to nonresponse because of an unwillingness or inability to participate in the survey interview. However, the cost associated with the use of proxy reporting *may* be an increase in the rate of errors of observation associated with poorer-quality reporting for others compared with the quality that would have been obtained under a rule of self-response.

The limited literature comparing self- and proxy reports in the measurement of disability has focused on the reporting of activities of daily living (Mathiowetz and Lair, 1994; Rodgers and Miller, 1997). Persons for whom data are obtained by proxy are often classified as having more functional limitations than those for whom the data are obtained by self-response; research is inconclusive as to whether this discrepancy is a function of overreporting on the part of proxy informants, underreporting on the part of self-respondents, or both.

Specific Wording of Question, Response Option Presentation, and Overall Context Although there is empirical evidence to indicate that estimates of the population differ as a result of different question wording, the presentation of response alternatives, the order of questions, and the overall context of the questionnaire, little is known concerning the measurement error properties of alternative approaches, nor is there empirical literature that addresses the marginal effects of various question design features.

As is evident from the design choices discussed above, each choice impacts the error structure of the estimates of persons with disabilities and the analytic capabilities that can be addressed with the resulting data. Also evident is the lack of information with respect to the specific impacts of design choices on the reporting of impairments and disabilities; this point was one of several made in the Workshop on the Measurement of Work Disability (Mathiowetz and Wunderlich, 2000).

One could consider a number of permutations of the options outlined above in designing a work disability measurement system; these options could be arrayed along lines of richness of the data, quality of the data, and costs. For example, consider a system with the following attributes:

- continuous, longitudinal multimode household-based data collection (so as to facilitate participation among those who are unable or unwilling to answer via a single mode);
- medical examination for those meeting a particular threshold based on the household data and a subset of those who are classified in the category adjacent to the threshold; and
- links to administrative records.

Such a design would facilitate the analysis of change over time in the size of the pool of eligibles and applicants, and the understanding of the individual and environmental factors that influence application for benefits, and could simulate the impact of alternative decision processes, provided the household survey, medical examination, and administrative records collected or contained the information necessary for such modeling. In contrast, one could consider a design that is characterized by a small number of questions on disability included as part of repeated cross-sectional surveys. Such a design would limit analysts to monitoring the size of the pool of eligibles and possibly, if cross-walk analytic capabilities had been developed, the size of the pool of applicants. However, such a design does not facilitate understanding how individual, environmental, and macrolevel changes impact the application process; with such a design one can only observe a correlation between macrolevel changes and changes in the size of the applicant pool, but cannot understand the relationship between the two at the individual level. Between these two extremes, one could consider a large number of variations.

Underlying the hypothetical continuum of design options is a second continuum related to the costs of alternative design combinations; the analytic capabilities associated with the richest design come at the cost of higher expenditures for data collection. Regardless of the choices made with respect to mode, frequency, cross-sectional versus longitudinal design, and other design features, further research to understand the error properties associated with alternative design features is necessary to more fully inform the decision process with respect to the cost–error trade-offs.

The choice of data collection design options is not one unique to the Social Security Administration. Other federal agencies responsible for constructing a social indicator series or providing data for the purposes of public policy or funds management have struggled with similar dilemmas. For example, the Agency for Healthcare Research and Quality (AHRQ, formerly the Agency for Health Care Policy and Research), faced a similar design issue with respect to the provision of information concerning health care utilization, expenditures, and health insurance coverage. During the 1970s and 1980s the agency relied on periodic household data collection efforts, supplemented with provider records and administrative data, as the basis for producing estimates (e.g., the 1977 National Medical Care Expenditure Survey, the 1986 National Medical Care Utilization and Expenditure Survey). These detailed data, yielding themselves to years of alternative analyses, formed the basis of long-range policy guidance. However rich these data were, such a design did facilitate research related to understanding shifts in health care utilization or expenditure patterns. In recent years, the design has shifted toward continuous data collection, with a longitudinal panel (the Medical Expenditure Panel Survey, see www.ahrq.gov/data/mepsinfo). The shift toward a longitudinal data collection effort with an ongoing (continuous) rotating panel design both increased the analytic capabilities of the data and reduced the gaps in data needed for public policy.

Partnerships with Other Federal Agencies

As noted above, one of the choices for SSA is whether to sponsor its own ongoing surveys or to enter into partnership with other federal agencies to obtain a small set of measures.

In short, what are the administrative, financial, and technical staffing burdens of mounting an ongoing survey, and what is the scope of informational needs? If there are many features of the population that are not now being well described, then a separate SSA survey may easily be justified as a small fraction of the funds allocated to fulfill its mission.

However, if a smaller set of measures would sufficiently measure the population of interest (both the pool of eligibles and the pool of applicants), as well as address the other analytic goals of SSA, then partnership with another federal agency or agencies may be a cost-effective option within the work disability measurement system.

The candidate surveys for ongoing monitoring include the American Community Survey, the American Housing Survey, the Behavioral Risk Factor Surveillance System, the Current Population Survey, the Medical Expenditure Panel Survey, the National Crime Victimization Survey, the National Health Interview Survey, the National Health and Nutrition Examination Survey, the National Household Survey of Drug Abuse, and the Survey of Income and Program Participation. Three criteria were used for selection of the candidate surveys discussed here: (1) each represents an ongoing federal data collection effort; (2) the sample size is sufficient, on an annual basis, to support SSA data requirements; and (3) the survey instrument currently includes or is planning to include measures of disability as part of the questionnaire. Some candidate surveys did not meet all three criteria but were included for consideration due to some unique design feature of the study. For example, the annual samples for the National Health and Nutrition Examination Survey and the Medical Expenditure Panel Survey (MEPS) are relatively small as compared to some other surveys ($n = 5,000$ and $n = 15,000$ persons annually, respectively); however, each of their designs benefits from a complementary component. In the case of the NHANES, the design includes a medical examination. In the case of the MEPS, the design includes data from medical care providers and providers of health insurance. Similarly, the National Household Survey of Drug Abuse does not presently include any measures of functional limitation or disability; however, the design includes both an interviewer-administered questionnaire and a self-administered set of questions that may be beneficial in the assessment of disability.

The relevant questions to be addressed in choosing a partner survey include the following:

- How large a sample is interviewed each year? What standard errors are likely to be obtained for key disability prevalence statistics?
- Will the addition of disability measures to the interview be consistent with the measurement goals of the original survey? Are there possibilities of context effects that could damage the accuracy of prevalence estimates?
- Are there existing measures in the survey that might be used as explanatory variables for disability status indicators? Can the survey offer SSA other informational benefits beyond being a vehicle to produce disability prevalence statistics?

- Is the survey of high quality? What evidence is there about coverage, nonresponse, and measurement error properties of key statistics?
- How frequently can estimates be updated? Will monthly prevalence estimates be generated, annual estimates, etc.?
- Is the mode of administration of the survey compatible with the measures chosen from NSHA?
- What restrictions, if any, will SSA staff have on access to microdata from the surveys? Can SSA analysts use the data for other analyses of importance to SSA or will they be given only statistics produced from the survey data?
- Will the mission of the sponsoring agency be aided by a partnership with SSA in measuring disability status? With the obligation of many federal household surveys to provide indicators of disability, can SSA expertise in work disability be viewed as a desirable complement to the sponsor's staff skills?

A partnership between two or more federal agencies may be beneficial to all parties involved. For example, collaborative efforts could lead to building a consensus concerning the measurement of disability in federal surveys. Additional funds from SSA to support data collection efforts may also support increases in sample size, further questionnaire development and refinement, and expand the analytic utility of any one data collection effort.

The ideal partner survey would have a sufficiently large sample[4] to provide SSA with prevalence estimates that were stable enough to protect policymakers from erroneous impressions. It would have very low coverage and nonresponse errors. It would be conducted frequently, giving SSA the ability to model seasonal effects in the size of the pool and to estimate the impact of economic shocks. It would contain other measures that would be of utility to SSA in addressing other important management problems: Are all demographic subgroups changing disability prevalence in the same way over time? What are the major health and demographic correlates of disability status?

The chief threat to the feasibility of this partnering option for ongoing monitoring is that most federal household surveys are already using long and complex instruments, filled with measures of great value to existing constituencies. Seeking to add measures to these instruments faces zero-sum conflicts with existing obligations of the sponsors. The single most important sign of optimism is that several of the surveys are facing man-

---

[4]"Sufficient" sample size could be based on the study's current sample or on a sample size resulting from the pooling of funds across agencies to gain efficiency in data collection.

dates to begin measurement of disability status in order to learn how the disabled subpopulation differs from others on the key topics covered by the surveys.

The discussion that follows outlines the characteristics of several large, ongoing federal data collection efforts, some of which do include measurement of impairments, functional limitations, and work limitations and disability. Each of these potential partner surveys has strengths and weaknesses that have to be assessed in light of the questions enumerated above.

Agency for Healthcare Research and Quality

Medical Expenditure Panel Survey The household component of the Medical Expenditure Panel Survey is designed as a continuous, overlapping panel design, in which members of each panel are interviewed for a two-year period concerning health care use, expenditures, sources of payment, and insurance coverage. Approximately 6,000 households are selected from those responding in the prior year to the National Health Interview Survey; household members are then interviewed five times over a 24-month field period, yielding information on approximately 15,000 persons for each panel. To produce estimates for any one particular calendar year, the data can be pooled across two distinct nationally representative samples, yielding an effective sample size of approximately 30,000 persons annually. The MEPS sample design targets for oversampling those with family income less than 200 percent of the poverty level, working age adults predicted to have high health care expenditures (based on information obtained in the NHIS interview), and adults 18 years of age and older classified as having a functional limitation, measured in terms of activities of daily living (ADLs) and instrumental activities of daily living (IADLs). In addition to the household panel survey, the MEPS design includes a survey of medical providers identified by MEPS respondents; data are collected from these medical providers to verify and supplement information provided by the household respondents. A second supplemental data collection involves contacting employers and other providers of health insurance identified by the household respondents so as to collect information on insurance characteristics that household respondents cannot usually provide.

Bureau of the Census

The U.S. Bureau of the Census conducts two surveys of interest, the American Community Survey and the Survey of Income and Program Participation.

American Community Survey The American Community Survey (ACS) is a new initiative of the Bureau of the Census, designed to eventually replace the long-form decennial census. The design of the survey closely resembles the decennial census, with self-administration of mail-delivered questionnaires. The sample consists of a rolling sample of addresses, with approximately 3 million households sampled annually. At present, the questions on disability replicate those included in the long form of the year 2000 decennial census. Drawing on the Canadian experience in conducting the Health and Activity Limitation Surveys (HALS), the ACS could be used as a first-stage screening instrument for the identification of individuals likely to be impaired or disabled; follow-up, in-depth interviews could be targeted at those individual identified via screening questions in the ACS as well as a subsample of those not identified as disabled, so as to capture the false negatives via the longer instrument.

Survey of Income and Program Participation SIPP is a multipanel longitudinal survey of adults, that measures their economic and demographic characteristics. Participants are interviewed once every 4 months; the duration of each panel ranges from 2.5 to 4 years. The questionnaire for the SIPP includes a core set of questions administered every wave and a set of topical modules, which are administered periodically. One of the topical modules that has been administered in previous panels concerns disability and functional limitations. The redesigned topical module administered in 1997 and 1999 covered a broad range of questions concerning disability and functional limitations, including sensory limitations, use of mobility aids, ADLs, IADLs, and upper- and lower-body functional limitations.

Bureau of Justice Statistics

National Crime Victimization Survey As a result of Public Law 105-301, the Bureau of Justice Statistics (BJS) is required to produce victimization rates by developmental disability status beginning in the year 2003. To meet this requirement, BJS has begun to develop and test a 20-question model dealing with health conditions, impairments, and disabilities, covering a broad range of disabilities, not just developmental disabilities. The questions have undergone testing in a cognitive laboratory and will be field-tested this spring among a population of persons with developmental disabilities in California. These questions would be added to the National Crime Victimization Survey (NCVS), a rotating panel design survey in which participants are interviewed every 6 months over a 3.5-year period. Similar to the design of the Current Population Survey, the sample unit for the NCVS is the housing unit; participants who move during the life of the panel are not followed. Approximately 50,000 households are interviewed every six months, with information collected on approximately 100,000 persons ages 12 and older annually.

Bureau of Labor Statistics

Current Population Survey The Current Population Survey (CPS) is a rotating panel design in which households are interviewed monthly for four months, not interviewed for eight months, and then interviewed monthly for an additional four months. The questionnaire consists of a core set of questions concerning labor force participation and, depending on the month of interview, a periodic or topical module. For example, detailed information concerning sources of income is collected for all participants who are interviewed during the month of March. The current CPS questionnaire obtains information concerning disability only when the respondent volunteers that he or she is disabled in response to the question concerning whether he or she worked last week for pay. In addition, respondents who are currently not employed are asked whether they have a disability that prevents them from accepting any kind of work during the next six months. Data are collected from approximately 60,000 households (on approximately 94,000 persons ages 16 and older) every month.

In response to Executive Order 13078, which requires the Bureau of Labor Statistics in conjunction with other federal agencies to produce accurate and reliable employment rate data for people with disabilities, the Bureau of Labor Statistics is evaluating a set of questions for possible inclusion in the CPS. About 20 questions were tested in cognitive labs and are currently being field-tested in the National Comorbidity Survey.

Centers for Disease Control and Prevention/National Center for Health Statistics

Behavioral Risk Factor Surveillance System The Behavioral Risk Factor Surveillance System (BRFSS) is a state-based surveillance system active in all 50 states and the District of Columbia. The data are collected by telephone, among adults ages 18 and older, on a monthly basis by individual states. Sample sizes vary by state and year but must be of sufficient size so as to permit state-level estimation for measures included in the core module. The BRFSS has three components: (1) a core questionnaire used in all states; (2) standardized modules chosen for inclusion by individual states; and (3) questions developed by each state. Beginning in the year 2000, the core module included the same two questions included in the National Health Interview Survey. One of the standardized supplemental modules

("Quality of Life") includes six questions on functional limitations and impairments. Disability measures can also be found in several additional standardized modules.

National Health Interview Survey The National Health Interview Survey is a cross-sectional survey conducted throughout the calendar year (nationally representative replicate samples are introduced every two weeks) that collects information about the amount and distribution of illness in terms of limited activities, chronic impairments, and health care services received by persons of all ages. All persons in the household are asked two questions concerning disability: (1) Are you limited in any way in any activities because of physical, mental, or emotional problems? (2) Do you now have any health problem that requires you to use special equipment, such as a cane, a wheelchair, a special bed, or a special telephone? Sampled adults (one per household) are asked a series of questions concerning functional limitations and the degree of difficulty associated with going out, participating in social activities, and participating in leisure activities in the home. Over the course of a year, data are collected on approximately 98,000 persons (core questionnaire), with additional information obtained from approximately 32,000 sampled adults and 14,000 sampled children.

National Health and Nutrition Examination Survey The redesigned National Health and Nutrition Examination Survey collects information on health and nutritional status of adults and children in the United States through household-based interviews as well as physical examinations. Although the NHANES was a periodic survey that began in the 1960s, in 1998 the study was redesigned so as to provide continuous monitoring of the population. The annual survey consists of interviews with approximately 5,000 persons per year. The household questionnaire includes questions concerning limitations in activities for children and for adults—limitations related to work, mobility, cognition, and functional activities. A medical examination also provides information on physical limitations as well as assessment of mental health and cognitive function.

Housing and Urban Development

American Housing Survey The American Housing Survey (AHS) consists of a national biannual sample and a rolling annual metropolitan sample conducted by the Census Bureau for the Department of Housing and Urban Development. National data are collected every other year from a fixed sample of housing units supplemented by a new construction sample. The national sample consists of approximately 55,000 housing units. In addition to the national sample, a metropolitan sample for each of 46 selected metropolitan areas is collected about every four years, with an average of 12 metropolitan areas included each year. Each metropolitan area sample covers approximately 4,800 or more housing units. The disability questions focus on questions related to mobility within the housing unit, limitations in activities of daily living, and sensory impairments.

Substance Abuse and Mental Health Services Administration

National Household Survey of Drug Abuse The National Household Survey of Drug Abuse (NHSDA) is a annual survey of approximately 67,000 persons concerning drug and alcohol use. The survey consists of an interviewer-administered as well as a self-administered section using computer-assisted interviewing techniques. The sample design consists of state-level cross-sectional samples, thereby facilitating state-level estimation. The current questionnaire does not include measures of functional limitations or disability.

DISCUSSION

Considering alternative design options for the ongoing measurement of persons with work disabilities requires careful consideration of alternative sources of error, the impact of various sources with respect to the estimates of interest, the analytic objectives of the data collection effort, and costs. As is evident from the preceding discussion, the empirical literature is, to a large extent, silent with respect to the impact of various sources of error on estimates of persons with work disabilities. Alternative designs will vary in the richness of the analytic capabilities of the resulting data; such capabilities will have to be balanced against issues of respondent burden and costs.

One issue is clear with respect to the design of an ongoing data collection effort to monitor the size of the applicant pool for SSA benefits. The lack of empirical data to inform the design at the present time emphasizes the need for SSA to undertake ongoing research as an integral part of the design of any data collection effort. An ongoing methodological research program, coupled with whatever design is implemented, will provide assessments of the error properties of the current design and inform future design decisions. The agenda for research in survey measurement outlined in the Institute of Medicine Workshop on the Measurement of Persons with Work Disabilities may provide a starting point for such a research effort.

REFERENCES

Altman B. 2001. Definitions of disability, and their operationalization, and measurement in survey data: An Update. *Research in Social Science and Disability* 2:77–100.

Bixby L. 1976. Retirement patterns in the United States: Research and policy interaction. *Social Security Bulletin* 3–19.

Bound J, Waidmann T. 2000. *Accounting for Recent Declines in Employment Rates Among the Working-Age Disabled.* Ann Arbor, MI: Population Studies Center, University of Michigan.

Bye B, Schechter E. 1982. *1978 Survey of Disability and Work.* SSA Publication No. 13-11745. Washington, DC: U.S. Department of Health and Human Services.

Haber L. 1973. Social planning for disability. *Journal of Human Resources* (February):33–55.

Jette A, Badley E. 2000. Conceptual issues in the measurement of work disability. In: Mathiowetz N, Wunderlich GS, eds. *Survey Measurement of Work Disability: Summary of a Workshop.* Washington, DC: National Academy Press.

Mathiowetz N. 2000. Methodological issues in the measurement of work disability. In: Mathiowetz N, Wunderlich GS, eds. *Survey Measurement of Work Disability: Summary of a Workshop.* Washington, DC: National Academy Press.

Mathiowetz N, Lair T. 1994. Getting better? Change or error in the measurement of functional limitations. *Journal of Economic and Social Measurement* 20:237–262.

Mathiowetz N, Wunderlich G, eds. 2000. *Survey Measurement of Work Disability: Summary of a Workshop.* Washington, DC: National Academy Press.

McNeil J. 1993. *Census Bureau Data on Persons with Disabilities: New Results and Old Questions about Validity and Reliability.* Paper presented at the 1993 Annual Meeting of the Society for Disability Studies, Seattle, Washington, 1993.

Nagi S. 1991. Disability concepts revisited: Implications for prevention. In: Pope A, Tarlov A, eds. *Disability in America: Toward a National Agenda for Prevention.* Washington, DC: National Academy Press.

Rodgers W, Miller B. 1997. A comparative analysis of ADL questions in surveys of older people. *Journal of Gerontology* 52B:21-36.

Sheppard H. 1977. Factors associated with early withdrawal from the labor force. In: Wolbein SL, ed. *Men in the Pre-Retirement Years.* Philadelphia: Temple University. Pp. 163–215.

World Health Organization. 2001. *International Classification of Functioning, Disability, and Health.* Geneva, Switzerland: World Health Organization.

Yelin EH, Nevitt MC, Epstein WV. 1980. Toward an epidemiology of work disability. *Milbank Memorial Fund Quarterly/Health and Society* 58(3):384-415.

Persons with Disabilities and Demands of the Contemporary Labor Market

Edward Yelin, Ph.D., and Laura Trupin, MPH[1]

Disability insurance programs, whether public or private, require an assessment of the ability of persons with disabilities to function in jobs. Although some of the problems inherent in such assessments—determining severity of illness, ascertaining physical and cognitive impairment— were noted early in the twentieth century with respect to private disability insurance programs and workers' compensation (Starr, 1982; Stone, 1984; Berkowitz, 1987; Derthick, 1990; Mashaw and Reno, 1996), some are new and reflect changes in the economy. The procedures that were implemented to assess work capacity in most disability insurance programs, including the Social Security Administration's (SSA's) Social Security Disability Insurance (SSDI) and Supplemental Security Income (SSI) programs, reflect an economy dominated by goods production, physical labor, hierarchical organization, and long job tenures (Yelin, 1992); a population thought to be at risk for work loss primarily because of the chronic diseases of aging (Chirikos, 1993; Stapleton, et al., 1994); and the view that most such conditions would lead, inexorably, to functional decline without the prospect for improvement.

This paper describes some of the changes in the labor market that have occurred over the last several decades, shows the extent to which the

---

[1]Edward Yelin is a Professor of Medicine and Health Policy and Director of the Arthritis Research Group at the University of California at San Francisco. Laura Trupin is a Senior Research Associate for the Institute for Health and Aging.

labor market experience of persons with disabilities reflects these trends, and speculates about the demands that are likely to be placed on workers in the next several decades.

LABOR MARKET DYNAMICS: 1960 TO THE PRESENT

Overview of Changes in the Labor Market

Although it would be hazardous to predict what the labor market will be like in the distant future, several of the most important trends have been unfolding for several decades and can be expected to continue in the years to come (Bell, 1973; Piore and Sabel, 1984; Hirshhorn, 1988; Levy, 1998; Wilson, 1997). These trends include a relative shift from goods-producing occupations and industries to the distribution of services, the increasing demand for highly skilled and highly trained labor and the erosion of demand for those with less skill and training, the emergence of new ways of accomplishing work within the firm, and the emergence of alternative work arrangements throughout the economy.

Some of these trends are relatively easy to quantify, for example, the growth of jobs in services. Some are more difficult both to measure and to evaluate, for example, the growth of contingent employment arrangements (Belous, 1989; Polivka, 1996), the putative erosion of job security (Nardone et al., 1997), and the flattening of workplace hierarchies (Osterman, 1988). Also, many of the changes are not quite as dramatic as some analysts claim: much service work is physically demanding and much of it, regardless of the physical demand, is repetitious. All, however, are difficult to translate into a simple set of instructions for assessing functional capacity for work. Indeed, if there is a message to emerge from an analysis of the trends in the labor market, it is that in the contemporary economy, the division of tasks within and among jobs is growing increasingly complex.

As work demands change, the most important characteristic of those capable of thriving may be the ability to do multiple tasks in an overlapping and constantly evolving series of relationships and to be able to adapt to new responsibilities. The problem facing those assessing capacity for work among persons with disabilities is a daunting one: how to assess an individual's capacity to do a complex mix of tasks now and to learn a new mix later.

Dynamics in Labor Force Participation

The 1950s and 1960s are viewed by some as the halcyon era in the U.S. economy, with high growth rates sustaining unprecedented increases in the standard of living, allowing most families to survive on one income, and in turn, reinforcing the social ethic of the time that women should not work outside the home (Levy, 1998). In 1960, 66.8 percent of the working-age population was in the labor force (Table 1). The overall labor force participation rate increased by more than 18 percent in the interim, reaching almost 80 percent as of 1999.

Gender

This overall increase in labor force participation rates masks substantial differences by gender and age. Among all working-age men, labor force participation rates declined by more than 7 percent, but men age 55 to 64 experienced an even steeper decline, just under 22 percent. Conversely, among all working-age women, labor force participation rates rose by 68.9 percent, from 42.7 percent in 1960 to 72.1 percent in 1999. Among women age 25 to 34, labor force participation rates more than doubled, from 36.0 percent in 1960 to 76.4 percent in 1999. Thus, the overall increase in labor force participation rates represents the net effect of a decline among men, particularly older men, and an increase among women, particularly younger women.

TABLE 1 Labor Force Participation Rates (percent), by Gender and Age, United States, 1960–1999

	Year						Percent Change, 1960–1999
Gender and Age	1960	1970	1980	1990	1996	1999	
	Percent						
All persons, 18–64	66.8	69.2	74.0	78.1	78.7	79.0	18.3
Men							
18–64	93.2	90.2	88.1	87.6	86.4	86.1	–7.6
55–64	86.8	83.0	72.1	67.8	67.0	67.9	–21.8
Women							
18–64	42.7	50.2	60.9	69.0	71.3	72.1	68.9
25–34	36.0	45.0	65.5	73.5	75.2	76.4	112.2

SOURCE: Jacobs and Zhang, 1998; U.S. Department of Labor, 1999a.

TABLE 2 Labor Force Participation Rate (percent), by Race and Gender, United States, 1972–1999

	Year					Percent Change, 1972–1999
Race and Gender	1972	1980	1990	1996	1999	
			Percent			
Whites	69.5	74.6	79.0	79.8	79.8	14.8
Men	90.1	89.1	88.7	87.8	87.5	–2.9
Women	50.4	60.8	69.5	71.9	72.2	43.3
African Americans	68.6	70.3	73.1	73.5	75.4	9.9
Men	83.8	80.9	80.5	77.9	77.8	–7.2
Women	56.1	61.7	67.1	69.8	73.4	30.8

SOURCE: Jacobs and Zhang, 1998; U.S. Department of Labor, 1999a.

Race

Race plays a part in labor market dynamics and would appear to interact with gender.[2] Over the last 27 years, labor force participation rates increased among all working-age whites by 14.8 percent, but the increase among all working-age African Americans was only 9.9 percent (Table 2). The decrease in labor force participation rates among all working-age white men was less than half that experienced by African-American men (2.9 versus 7.2 percent, respectively), while the increase among white women was far larger than that among African-American women (43.3 versus 30.8 percent, respectively). Between 1972 and 1999, the gap in labor force participation rates between African-American and white men grew, from 6.3 percentage points in the former year to 9.7 percentage points in the latter. In 1972, labor force participation rates of African-American women were higher than those of white women (56.1 and 50.4 percent, respectively), but by 1999 the groups had virtually identical labor force participation rates (73.4 and 72.2 percent, respectively).

Age

Another factor affecting the labor market over the last several decades—and one likely to have an even more profound impact in the years

[2]Prior to 1972, published labor market series combined all non-Caucasians into one category. Accordingly, we report racial differences in labor force participation from 1972 to 1999.

to come—has been the dramatic change in the age structure of society as the baby boomers age (Table 3). The proportion of the population 18 to 34 years of age rose substantially between 1960 and 1980 but has since fallen, while the proportion 35 to 44 rose between 1980 and 1999, and the proportion 45 to 54 began a precipitous increase during the 1990s, to be followed in the decade to come by a substantial rise in the proportion of individuals 55 and over.

The importance of the aging of the population for the labor market can be seen in Table 4. In 1999, more than 80 percent of people 20 to 34, 35 to 44, and 45 to 54 years of age were in the labor force. In each case, these percentages had risen over time as the labor market accommodated the substantial increases in labor force participation rates among women. The increase in the labor force participation rates were all the more remark-

TABLE 3 Age Structure (percent) of United States Population, 1960–1999

Age	Year					
	1960	1970	1980	1990	1996	1999
	Percent					
18–34	21.6	24.4	29.6	28.2	24.5	23.5
35–44	13.4	11.3	11.3	15.1	16.4	16.4
45–54	11.4	11.4	10.1	10.1	12.2	13.1
55–64	8.6	9.1	9.6	8.5	8.1	8.6
65 or older	9.2	9.8	11.3	12.5	12.8	12.7

SOURCE: U.S. Bureau of the Census, 1984, 1997, 2000.

TABLE 4 Labor Force Participation Rates (percent), by Age, United States, 1960–1999

Age	Year					
	1960	1970	1980	1990	1996	1999
	Percent					
20–34	65.3	69.5	78.9	81.8	81.9	82.4
35–44	69.4	73.1	80.0	85.2	84.8	84.9
45–54	72.1	73.5	74.9	80.7	82.1	82.6
55–64	60.9	61.8	55.7	55.9	57.9	59.3
65 or older	20.8	17.0	12.5	11.8	12.1	12.3

SOURCE: Jacobs and Zhang, 1998; U.S. Department of Labor, 1999a.

able given that the absolute number of young and middle-aged workers was increasing because of the baby boom generation. Thus, the labor market accommodated an increasing percentage of a substantially larger number of persons.

However, labor force participation rates are much lower among persons age 55 to 64 than among those age 45 to 54, and they declined among persons in the former group throughout most of the 1970s and 1980s. The decrease in labor force participation rates among persons age 55 to 64 before 1990 occurred because more people in this group chose to leave work prior to the ages when Social Security eligibility begins (62) and reaches its maximum (currently 65). Labor force participation rates are lower among persons age 55 to 64 at any one point because persons in this group face higher rates of displacement from their jobs and because the prevalence of health problems associated with aging begin to affect a substantial number of people at these ages. As a result of the increased number of persons who are 55 to 64, a higher proportion of the working-age population will be at risk for onset of the chronic diseases of aging, putting increased pressure on disability compensation programs. On the other hand, among persons age 55 to 64, labor force participation rates have increased over the last decade, suggesting that a strong labor market affects the propensity of persons in this group to leave the labor force.

Education

As seen in Table 1, the proportion of working-age adults in the labor force rose substantially between 1970 and 1999. The increase in labor force participation rates affected all but those individuals who did not finished high school (Table 5). Thus, labor force participation rates increased among high school graduates by 11.3 percent, among those with some college by 12.5 percent, and among those with a college degree or more, by 6.4 percent. As a result, by 1999, labor force participation rates among college graduates were 40 percent higher than among persons with less than a high school education.

Since 1960, the proportion of the adult population with at least a high school degree has more than doubled (from 41.1 to 83.4 percent), and the proportion with four or more years of college has more than tripled (from 7.7 to 25.2 percent) (U.S. Bureau of the Census, 2000, p. 157). Nevertheless, a substantial fraction of the cohorts entering the ages of highest risk for work disability have less than a high school education, including about 12 percent of those now ages 35 to 44 and 45 to 54, and more than 18 percent of those now age 55 to 64 (U.S. Bureau of the Census, 2000, p. 158). These individuals may face a difficult time maintaining a toehold in the labor market. In addition, about a third of these cohorts (33.9, 31.7, and 36.9

TABLE 5 Labor Force Participation Rates (percent), by Educational Attainment, United States, 1970–1999

Education	Year					Percent Change, 1970–1999
	1970	1980	1990	1996	1999	
			Percent			
Less than high school	65.5	60.7	60.7	60.2	62.7	–4.3
High school graduate	70.2	74.2	78.2	77.9	78.1	11.3
Some college	73.8	79.5	83.3	83.7	83.0	12.5
College grad or more	82.3	86.1	88.4	87.8	87.6	6.4
			Gradient[a]			
	1.26	1.42	1.46	1.46	1.40	

[a]Gradient from highest to lowest level of education.
SOURCE: U.S. Bureau of the Census, 1997, 2000.

percent, respectively) have no more than a high school degree. Although the labor force participation rate for high school graduates has increased by 11.3 percent overall since 1970, it has been relatively stable since 1990. If the labor market continues to tighten in the next few years, labor force participation rates among high school graduates may begin to fall.

Dynamics in Employment Characteristics

There is little doubt that there has been a fundamental shift in the *kind* of work done, as reflected in the change in the distribution of occupations and industries. However, analysts disagree on the degree to which there has been a corresponding shift in *how* work is done. Osterman (1988) noted that throughout much of this century, firms had two kinds of employees: a salaried workforce paid to design and monitor work processes, who were given relative autonomy to carry out their work and had security of employment ("white-collar" workers), and an hourly wage workforce paid to implement these work processes with little discretion over how work was done, who were retained only when the demand for products justified continued employment ("blue-collar" workers). Osterman observed that more recently, many firms were melding the two kinds of jobs: bringing the expertise of those involved in production of goods and services into the design of work processes, while reducing the security of employment among the white-collar workforce.

The signposts for the changes described by Osterman include flattened workplace hierarchies, broadened and variable work tasks for each job, reduced job tenure, increased use of part-time and temporary workers, alternative work arrangements, and higher rates of job displacement. There is strong evidence in the work disability literature that providing flexible working conditions and job autonomy reduces the probability that an individual with an impairment will stop working (Yelin et al., 1980). Indeed, the Americans with Disabilities Act of 1990 (ADA) mandates the provision of such accommodations to help sustain employment (West, 1991). The model underlying research on the effects of accommodation on employment, as well as the reasonable accommodation provisions of the ADA, is that increased autonomy to perform an existing mix of job demands in the context of a long-term relationship with an employer will improve job prospects. However, it is not known how well persons with disabilities can function when asked to flexibly shift among job tasks and work groups, especially with decreased levels of job security.

Ongoing data collection efforts at the Department of Labor's (DOL's) Bureau of Labor Statistics (BLS) measure some of the shifts in working conditions—job tenure, frequency of part-time and temporary employment, alternative work arrangements, and rates of job displacement. They do not capture changes in the nature of workplace hierarchies and in the mix of work tasks for each job. Obtaining such information will be critical in assessing the functional demands of work and, therefore, in assessing the capacity of persons with disabilities to function on the job.

Industries

Table 6 shows the change in the number of employees and share of nonagricultural employment among industries since 1960. It provides information on the most tangible signpost of the change in the nature of work. In 1960, the goods-producing sectors of the economy (mining and construction, and manufacturing) accounted for 6.7 and 31.0 percent of employment, respectively. Since then, the share of employment accounted for by mining and construction has decreased by about one-fifth, and the share accounted for by manufacturing has decreased by more than half (53.9 percent). Indeed, at a time when total employment more than doubled (datum not in table), the absolute number of manufacturing workers increased by only 9.5 percent, from 16.8 million in 1960 to 18.4 million in 1999. Thus, as of 1999, the goods-producing sectors of the economy accounted for less than one-fifth of total employment.

Concurrently, there was substantial growth in the share of employment accounted for by the finance, insurance, and real estate sectors (20.4 percent, net of a decline from 6.1 to 5.9 percent between 1990 and 1999)

TABLE 6 Number of Employees and Shares of Nonagricultural Employment, by Industry, United States, 1960–1999

Industry	Year						Percent Change, 1960–1999
	1960	1970	1980	1990	1996	1999	
	Numbers (millions)						
Mining and construction	3.6	4.2	5.4	5.8	6.0	6.8	88.9
Manufacturing	16.8	19.4	20.0	19.1	18.2	18.4	9.5
Transportation, utilities, and communications	4.0	4.5	5.2	5.8	6.4	6.8	70.0
Wholesale and retail trade	11.4	15.0	20.3	25.8	28.2	29.8	161.4
Finance, insurance, and real estate	2.6	3.7	5.2	6.7	7.0	7.6	192.3
Services	7.4	11.6	17.9	27.9	34.4	39.0	427.0
Public administration	8.4	12.6	16.2	18.3	19.5	20.2	140.5
	Percent in Nonagricultural Employment						
Mining and construction	6.7	6.0	5.9	5.3	5.0	5.3	−20.9
Manufacturing	31.0	27.3	22.4	17.4	15.3	14.3	−53.9
Transportation, utilities, and communications	7.4	6.4	5.7	5.3	5.3	5.3	−28.4
Wholesale and retail trade	21.0	21.3	22.5	23.5	23.6	23.1	10.0
Finance, insurance, and real estate	4.9	5.1	5.7	6.1	5.8	5.9	20.4
Services	13.6	16.3	19.8	25.5	28.7	30.3	122.8
Public administration	15.4	17.7	18.0	16.7	16.3	15.7	1.9[a]

[a]Percent change from 1980 to 1999 = −12.8%.
SOURCE: U.S. Bureau of the Census, 1981, 1997, 2000.

and by the service industry (122.8 percent). Primarily because of growth occurring prior to 1980, the share of total employment accounted for by the public administration sector increased by 1.9 percent; since 1980, however, its share has declined by 12.8 percent.

Because the service sector is heterogeneous, encompassing, for example, those who work in private households, physicians' offices, engineering firms, and home cleaning services, it is far more informative to study the employment dynamics within the components of the overall services category. The share of employment in all but the personal services component expanded between 1970 and 1999, with business and repair, entertainment and recreation, and professional services growing by 263.2, 100.0, and 44.8 percent, respectively (Table 7). By 1999, the absolute number of workers in professional services exceeded 32 million, almost a quarter of all non-farm employment. Within the business and repair services component, the absolute number of workers in personnel supply firms (including temporary employment agencies) increased more than fourfold between 1980 and 1999, while the number in the computer and data processing services fields increased more than ninefold (data on personnel supply and computer and data processing fields not in table) (U.S. Bureau of the Census, 2000, p. 420).

Occupations

The change in the share of employment among occupations reflects the shift in the overall economy from the production of goods to the production and distribution of services (Table 8). Thus, the share of employment in professional specialty and managerial occupations; techni-

TABLE 7 Number of Employees and Shares of Nonagricultural Employment in Various Service Industries, United States, 1970–1999

Service Industry	1970	1980	1990	1996	1999	Percent Change, 1970–1999
	Number (millions)					
Business and repair	1.4	3.9	7.5	8.1	9.0	542.9
Personal	4.3	3.8	4.7	4.4	4.5	4.7
Entertainment and						
recreation	0.7	1.1	1.5	2.4	2.6	271.4
Professional	12.9	19.9	25.4	30.1	32.4	151.2
	Percent in Nonagricultural Employment					
Business and repair	1.9	4.0	6.5	6.6	6.9	263.2
Personal	5.7	4.0	4.1	3.5	3.4	−40.4
Entertainment and						
recreation	1.0	1.1	1.3	1.9	2.0	100.0
Professional	17.2	20.7	21.9	24.4	24.9	44.8

SOURCE: U.S. Bureau of the Census, 1997, 2000.

TABLE 8 Number of Employees and Shares of Employment, by Occupation, United States, 1960–1999

Occupation	Year						Percent Change, 1960–1999
	1960	1970	1980	1990	1996	1999	
	Numbers (millions)						
Professional specialty and managerial occupations	14.6	19.4	26.5	30.6	36.5	40.5	177.4
Technical, sales, and administrative workers	14.0	18.6	24.3	36.9	37.7	38.9	177.9
Service workers	8.0	9.7	13.0	16.0	17.2	17.9	123.8
Precision production and craft workers	8.6	10.2	12.5	13.7	13.6	14.6	69.8
Operatives, fabricators, and non-farm laborers	15.6	17.6	18.4	18.2	18.2	18.2	16.7
Farming and fishing occupations	5.2	3.3	2.7	3.5	3.6	3.4	−34.6
	Percent Share of Employment						
Professional specialty and managerial occupations	22.1	24.7	27.3	25.8	28.8	30.3	37.1
Technical, sales, and administrative workers	21.3	23.6	25.0	31.1	29.7	29.2	37.1
Service workers	12.2	12.4	13.3	13.5	13.6	13.4	9.8
Precision production and craft workers	13.0	12.9	12.9	11.6	10.7	10.9	−16.2
Operatives, fabricators, and non-farm laborers	23.6	22.4	18.9	15.2	14.4	13.6	−42.4
Farming and fishing occupations	7.8	4.0	2.8	2.9	2.8	2.6	−66.7

SOURCE: U.S. Bureau of the Census, 1981, 1997, 2000.

cal, sales, and administrative workers; and service workers increased by 37.1, 37.1, and 9.8 percent, respectively, while the share in precision production and craft occupations; operatives, fabricators, and non-farm laborers; and farming and fishing occupations decreased by 16.2, 42.4, and 66.7 percent, respectively.

The shift from manufacturing to service occupations does not necessarily mean an absolute reduction in the former. Indeed, in absolute terms, the number of precision production and craft workers and operatives, fabricators, and non-farm laborers has increased by more than 8 million since 1960, although it has been relatively stable since 1980. Among major occupational classifications, only farming and fishing occupations have declined in absolute terms throughout the period covered. In contrast, the absolute number of persons in professional and managerial and technical, sales, and administrative occupations has more than doubled (from less than 14.6 million to 40.5 million in the former and from 14.0 million to 38.9 million in the latter). The number of service workers has also increased more than twofold (from 8.0 million to 17.9 million). Growth in the number of professional and managerial workers has continued throughout the period, with a particularly rapid increase in the number of workers in this group of occupations during the 1990s. Growth among technical, sales, and administrative and service workers has slowed since 1990. The recent rapid growth in professional and managerial occupations and the concurrent stasis among technical, sales, and administrative and service workers belie the prediction that the American economy would be producing few good jobs and many bad ones (Braverman, 1974; Wright and Singleman, 1982).

Part-Time Employment

The proportion of the employed population working part-time has increased since 1970, from 13.2 to 16.6 percent, or by more than 25 percent in relative terms, although it decreased during the late 1990s (Table 9). BLS divides part-time employment into voluntary and involuntary components (labeled "noneconomic" and "economic" reasons, respectively). The proportion of all employment that is part-time due to economic reasons increased from 2.8 to 4.3 percent between 1970 and 1990, but decreased to 2.5 percent as of 1999, because of the improvement in the labor market. In contrast, the proportion of the total employed population working part-time for noneconomic reasons continued to increase, having grown by more than a third since 1970, from 10.4 to 14.1 percent of the employed population.

TABLE 9 Percentage of Jobholders Working Part-Time for Economic, Noneconomic, and All Reasons, United States, 1970–1999

Reason	Year					Percent Change, 1970–1999
	1970	1980	1990	1996	1999	
			Percent			
All	13.2	15.1	17.2	17.4	16.6	25.8
Economic	2.8	4.1	4.3	3.4	2.5	−10.7
Noneconomic	10.4	11.0	12.9	14.0	14.1	35.6

SOURCE: U.S. Department of Labor, 1985, 1988, 2001; U.S. Bureau of the Census, 1990.

Terms of Employment

It is frequently claimed that an increasing fraction of all work is not in the traditional mode of being permanent, reasonably secure, in the direct employ of the firm in which the work is done, and with the work done at a worksite maintained by the firm. The Bureau of Labor Statistics has kept abreast of many of the changes in the terms of employment in its data collection efforts, but trend data are not available for many of them.

Job Security Job security is measured by length of time on the job (tenure) and the expectation of staying on the same job for an additional year (contingency) (Nardone et al., 1997). Among men, the overall median job tenure has not changed much since the early 1980s because the male workforce has aged and older workers have longer tenures. Within each age range, job tenure among men has decreased. Among women, job tenure has increased both because the fraction in older age groups has increased and because tenure for women 35 to 44 and 45 to 55 years of age has increased (U.S. Department of Labor, 1997). Thus, the picture for job tenure is a mixed one, with women having unambiguously longer tenures and men having shorter tenures at each age, but with more men being of the ages in which job tenures tend to be longer. Interestingly, job tenure has been falling for both men and women since 1996, suggesting that the strong labor market in the late 1990s may have resulted in shorter tenures as workers left old jobs for new ones and those who had been out of work found jobs (U.S. Department of Labor, 2000).

Contingent Employment BLS defines *contingent employment* in three ways: (1) the proportion of wage and salary workers whose jobs have lasted a year or more but who do not expect them to last another year; (2) the proportion of such workers as well as the self-employed and independent contractors in this situation; and (3) the proportion of both groups who do not expect their jobs to last another year regardless of how long they have been in them. The proportion meeting each definition declined slightly between 1995 and 1999. For the first definition, the decrease was from 2.2 to 1.9 percent of all workers; for the second the decrease was from 2.8 to 2.3 percent; and for the third the decrease was from 4.9 to 4.4 percent (U.S. Department of Labor, 1999a). Thus, contingency is reasonably common but has definitely not increased in the last few years. It should be reemphasized, however, that the recent decline may be due to the strength of the labor market in the last few years and may not reflect a long-term trend in the security of employment.

Alternative Work Arrangements Alternative work arrangements involve the shift from the direct hiring of workers to perform certain functions to the purchase of the services of other firms for those functions. These include the use of independent contractors, on-call workers, workers provided by temporary help agencies, and workers provided by contract firms. BLS has collected information on such arrangements only three times: in 1995, 1997, and 1999. The proportion of the employed with alternative work arrangements did not change substantially during this four-year period. As of 1999, 6.3 percent of all workers were independent contractors, 1.5 percent were on-call workers, 0.9 percent worked for temporary help agencies, and 0.6 percent were workers provided by contract firms (U.S. Department of Labor, 1999b).

Procurement of services outside the firm does not necessarily reduce the number of employees in the firm because outside services may be new or firm employees may be shifted to new functions as their old functions are outsourced. BLS collects information on proxy measures of the magnitude of employment in industries and occupations that represent services that could be done outside a firm (Clinton, 1997). Data on such measures suggest substantial growth in procurement of services outside firms. The share of total employment in the business services sector has increased threefold since 1972, and one component of this industry, personnel supply, has increased more than sevenfold during this time. In addition, there has been substantial growth in the engineering and management consulting sectors. Also, firms in a majority of industries have reduced their direct employment of business support occupations, those occupations that are most likely to be performed by outside contractors.

Change in Location of Work BLS collected information on the number of persons who do at least part of their jobs from home in 1991 and 1997 (U.S. Department of Labor, 1998). The number of persons who do some work at home was slightly more than 21 million (17.8 percent of the workforce) in 1997 and has not increased substantially since 1991. Almost two-thirds of persons who work at home are in managerial and professional specialty occupations.

Change in the Internal Structure of Work The workplace literature suggests a trend to diffuse authority over decisions about the way work is done throughout the hierarchy, to increased use of flexible work groups that coalesce only for the duration of specific projects, and to an increase in the mix of tasks done by the individual (Cornfield, 1987; Osterman, 1988; Kelley, 1990; Hirschhorn, 1991). The evidence for this kind of shift derives from qualitative studies of work settings (such as the shop floor and office) and from interviews and case studies of managers and line workers. However, without quantitative evidence, it is difficult to ascertain what proportion of the workforce has experienced these changes. In the 1970s, the DOL collected this kind of data in the Quality of Employment surveys; it has not been collected since (Quinn and Staines, 1979; Schwartz et al., 1988).

The potential importance of changes in the internal organization of work for persons with disabilities is profound. Flexibility in the pace and schedule of work and autonomy in how work is done have been shown to be strongly correlated with whether or not someone is able to maintain employment (Yelin et al., 1980). Thus, if the observation that these conditions are more prevalent in work now than in the past was true, it might augur an improvement in the employment picture for persons with disabilities. On the other hand, for persons with cognitive, communications, and psychological disabilities, the need to interact with a constantly changing array of workgroups and the impermanent working conditions may make it more difficult to work. Although it would be hard to capture these qualitative changes in working conditions in large-scale labor market surveys, they may be more important in determining the employment prospects of persons with disabilities than the more objective changes in employment described above.

Rates of Displacement BLS defines job displacement as the loss of a job held on a long-term basis (three or more years). BLS has tracked job displacement since the early 1980s (Hipple, 1997, 1999). The overall rate of job displacement seems tied to the economic cycle. It rose with the recession in the early 1980s, fell with the recovery late in that decade, rose once again with the recession early in this decade, and has since fallen. However, the composition of displaced workers has changed considerably. In the early years of the BLS data collection efforts, the rate of displacement was greater in manufacturing industries and in occupations such as craft workers and operatives that were concentrated in those industries. In the interim, the rate of displacement has grown faster in white-collar occupations and is now almost as great in such occupations as in blue-collar ones. It has also begun to spread to rapidly expanding industries, such as the finance, insurance, and real estate sectors. Thus, although a large proportion of displacement is due to cyclical changes in the economy, a portion of job displacement occurs in successful and expanding sectors. Job displacement is becoming a more generalized strategy of accommodating change in the labor force and is not limited to select occupations and industries facing difficult times.

The Labor Market and Persons with Disabilities

Persons with disabilities have experienced most of the major trends in the labor market over the last several decades, albeit in exaggerated form. In this section, we review the evidence to support this statement. The data on time trends among persons with disabilities, however, do not cover the same periods as the general labor market data reviewed in the prior section because most federal data series do not collect information on disability status with the same regularity as they do characteristics such as gender, race, and age.

Another factor affecting the study of labor market trends among persons with disabilities is the lack of a consistent definition of disability by the various data series. The National Health Interview Survey (NHIS), for example, defines disability as a limitation in a major life activity, such as work, housework, or school or, more broadly, as any limitation in any activity. Under this latter definition, approximately 14.1 percent of working-age adults were considered to have a disability in 1995 (Benson and Marano, 1998). By contrast, the Current Population Survey (CPS) measures only limitations in work, which reduced the prevalence of disability in the working-age population to about 8.0 percent in that year.[3] The CPS disability measure no doubt captures the severe end of the impairment spectrum, thereby artifactually reducing estimates of labor force participation rates among persons with disabilities. The impact of the different definitions of disability on estimates of labor force participation has recently become a topic of discussion in the disability literature (Hale, 2001). The reader is advised to note the data source for each table when drawing conclusions about the results presented.

---

[3]Authors' analysis of 1995 CPS.

TABLE 10 Labor Force Participation Rates (percent) of Persons with and Without Disabilities, by Gender, United States, 1983–1994

	Year		Percent Change
Gender and Disability Status	1983	1994	
	Percent		
All persons	75.0	78.6	4.8
With disabilities	48.6	51.8	6.6
Without disabilities	79.1	83.0	4.9
All men	87.2	86.9	–0.3
With disabilities	60.0	58.8	–2.0
Without disabilities	91.5	91.4	–0.1
All women	63.8	70.6	10.7
With disabilities	38.0	45.6	20.0
Without disabilities	67.6	74.9	10.8

SOURCE: Adapted from Trupin et al., 1997.

Labor Force Participation Rates

Between 1983 and 1994, labor force participation rates among all working-age persons increased by 4.8 percent (Table 10).[4] Although persons with disabilities continue to have lower labor force participation rates than persons without disabilities (51.8 and 83.0 percent, respectively), such persons experienced a larger relative increase (6.6 percent) than those without (4.9 percent). Thus, persons with disabilities more than shared in the overall increase in the proportion of working-age adults actually in the labor force. Several studies using data from sources other than the NHIS have recently been published. The results of the studies are not consistent (Bound and Waidmann, 2000; Burkhauser et al., 2000; Levine, 2000; McNeil, 2001) and have been criticized as not having adequately measured disability (Hale, 2001).

Gender, Age, and Race

Persons with disabilities experienced trends in labor force participation by gender to a heightened degree (Table 10). Thus, while labor force

---

[4]Unless otherwise noted, results presented here regarding persons with disabilities use the NHIS definition, based on overall activity limitations.

participation rates were increasing 10.8 percent among women without disabilities between 1983 and 1994, women with disabilities experienced an increase of almost twice that magnitude during this time (20.0 percent). Concurrently, men with disabilities experienced a larger decline in labor force participation rates than men without (2.0 and 0.1 percent, respectively).

Recall from Tables 2 and 3, that the decline in labor force participation rates among men was concentrated among those age 55 to 64, particularly nonwhite men in this age range, and that the increase in labor force participation rates among women was concentrated among women age 25 to 34, especially white women in this age range. Persons with disabilities experienced each of these trends in a heightened form (Yelin and Katz, 1994). Thus, labor force participation rates among men age 55 to 64 with disabilities declined to a greater degree than those among such men without disabilities, and nonwhite men of this age with disabilities experienced the largest relative decline in labor force participation of any single group defined by gender, age, race, and disability status. In contrast, young women with disabilities, particularly young white women, experienced the largest increase of any single group defined by these four characteristics.

Education

Persons with disabilities are overrepresented among persons with a high school education or less and underrepresented among those with some college or more (data from authors' analysis of 2000 CPS).[5] However, at every level of education, they have lower labor force participation rates than persons without disabilities, even after statistical adjustment for differences in demographic characteristics (Table 11). The difference in labor force participation rates is greater at lower levels of education. For example, the labor force participation rate among persons with disabilities with less than a high school education is about one-fifth as great as that among such persons without disabilities (14.5 and 73.6 percent, respectively), but persons with disabilities who have some graduate school or more have a labor force participation rate more than half that of persons without disabilities (48.9 and 89.8 percent, respectively). Attaining higher levels of education improves the employment prospects of persons with disabilities to a greater degree than persons without dis-

---

[5]The analyses presented in the sections on education, industries, occupation, and part-time employment derive from the CPS and use a measure of disability, therefore, that is based on work limitations only.

TABLE 11 Labor Force Participation Rate (percent) of Persons with and Without Disabilities, by Educational Attainment, with Adjustment for Demographic Characteristics, United States, 1999

	Labor Force Participation	
Educational Attainment	With Disabilities	Without Disabilities
	Percent	
Less than high school	14.5	73.6
High school	27.6	83.2
Some college	32.9	83.8
College graduate	42.5	87.0
Some graduate school or more	48.9	89.8

SOURCE: Authors' analyses of Current Population Survey 2000 Annual Demographic Supplement.

abilities, but even when persons with disabilities have gone to graduate school, they still have lower labor force participation rates than persons without disabilities who have not completed high school.

Employment Characteristics and Persons with Disabilities

Given employment, do persons with disabilities have access to the same mix of jobs and to the same working conditions as those without disabilities?

Industries

Recall from Table 6 that three industrial sectors have had a declining share of employment (mining and construction; manufacturing; and transportation, utilities, and communications); three have had a substantially increasing share (wholesale/retail trade; finance, insurance, and real estate; and services); and one has had little change, net of an increase prior to 1980 and a decline since then (public administration). Table 12 shows the mix of industries in 1999 among persons with and without disabilities who were employed. There are no clear patterns. Persons with disabilities are underrepresented in one sector with a declining share of employment (manufacturing) and in one with an increasing share (finance, insurance, and real estate), but they have a larger share of overall employment in the service industry and in two of the components of this sector—business and repair, and personal services. Persons with disabilities have a slightly smaller share of employment in professional services than persons with-

TABLE 12 Shares of Employment (percent) of Persons with and Without Disabilities by Industry, United States, 1999

Industry	Persons Employed		
	With Disability	Without Disability	Ratio
	Percent		
Mining and construction	6.8	7.6	0.90
Manufacturing	12.5	16.1	0.78
Transportation, communications, and utilities	7.5	7.5	1.00
Wholesale/retail trade	24.4	20.5	1.19
Finance, insurance, and real estate	4.6	6.7	0.69
Services	39.3	37.1	1.06
Business and repair	9.6	7.4	1.30
Personal	3.8	2.8	1.39
Entertainment and recreation	1.6	1.9	0.87
Professional	24.3	25.1	0.97
Public administration	4.9	4.6	1.05

SOURCE: Authors' analyses of Current Population Survey 2000 Annual Demographic Supplement.

out disabilities, the largest service industry component. They have an equal share of employment in transportation, utilities, and communications industries as persons without disabilities.

Occupations

The occupations with an increased share of employment over the last several decades include professional and managerial occupations; technical, sales, and administrative workers; and service occupations, while craft workers, operatives, and farming and fishing occupations have had declining shares of employment. With respect to occupations with an increased share of employment, persons with disabilities are much less likely than those without to be in professional and managerial occupations; they are almost as likely to be in technical, sales, and administrative occupations; and they are more likely to be service workers (Table 13). With respect to occupations with a declining share of employment, persons with disabilities are slightly less likely than those without to be in the precision production and craft trades, but persons with disabilities are substantially more likely to be operatives and to be in farming and fishing occupations.

TABLE 13 Shares of Employment (percent) of Persons with and Without Disabilities, by Occupation, United States, 1999

	Persons Employed		
Occupation	With Disabilities	Without Disabilities	Ratio
	Percent		
Professional specialty and managerial occupations	21.3	31.0	0.69
Technical, sales, and administrative workers	27.6	29.2	0.95
Service workers	20.1	13.2	1.52
Precision production and craft workers	10.0	11.1	0.90
Operatives, fabricators, and non-farm laborers	17.7	13.4	1.32
Farming and fishing occupations	3.4	2.2	1.55

SOURCE: Authors' analyses of Current Population Survey 2000 Annual Demographic Supplement.

Part-Time Employment

Persons with disabilities have experienced a disproportionate amount of the increase in part-time employment (Table 14). As of 1999, persons with disabilities reported that 36.0 percent of their employment was part-time, an increase of 29.0 percent since 1981. Concurrently, persons without disabilities experienced a 9.0 percent decrease in the percentage of part-time employment, from 16.7 percent in 1981 to 15.2 percent in 1999. Among persons with disabilities, the prevalence of part-time work due to economic reasons rose at least until the early 1990s, fell between 1990 and 1995, and has risen slightly in the interim, yielding a net decline of 12.7 percent over the entire period. Among persons without disabilities, part-time employment for economic reasons has fallen steadily since the mid-1980s, or by 41.9 percent overall between 1981 and 1999.

Persons with disabilities experienced a substantial increase in part-time employment for noneconomic reasons during the early part of the 1990s, leading to an overall increase of 41.2 percent in this measure over the entire period under study. In contrast, the rate of part-time employment for noneconomic reasons has not changed much among those without disabilities since 1981, having risen overall by only 2.4 percent in relative terms.

TABLE 14 Percentage of Jobholders Working Part-Time for Economic, Noneconomic, and All Reasons, Among Persons with and Without Disabilities, United States, 1981–1999

Reason	Year					Percent Change, 1981–1999
	1981	1985	1990	1995	1999	
	Percent					
All reasons						
Persons with disabilities	27.9	28.2	33.8	36.9	36.0	29.0
Persons without disabilities	16.7	17.1	16.5	16.7	15.2	–9.0
Economic						
Persons with disabilities	6.3	7.9	9.1	5.0	5.5	–12.7
Persons without disabilities	4.3	5.2	4.1	3.6	2.5	–41.9
Noneconomic						
Persons with disabilities	21.6	20.3	24.7	31.9	30.5	41.2
Persons without disabilities	12.4	11.9	12.4	13.1	12.7	2.4

SOURCE: Authors' analyses of Current Population Survey Annual Demographic Supplements for 1982, 1986, 1991, 1996, and 2000.

Terms of Employment

Of the measures of the terms of employment reviewed with respect to the entire labor force, above, none is available on an ongoing basis from the monthly CPS or the annual march supplement to the CPS. Instead, the measures—tenure, contingency, flexibility, alternative work arrangements, and work at home—are not collected routinely and, when collected, are part of surveys in which respondents are not asked to report disability status. Because of the lack of consistent data on terms of employment among persons with and without disabilities from the Bureau of Labor Statistics surveys, we report here the results of a comprehensive survey of health and employment among California adults, the 1999 California Work and Health Survey (Table 15).

In general, persons with disabilities did not differ systematically from those without in the working conditions they reported. On an unadjusted basis, persons with disabilities were more likely to report working at home. After adjustment for differences in age and gender, persons with disabilities reported significantly shorter job tenures and were significantly more likely to report holding their jobs for only one or five years than persons without disabilities. Of note, the two groups did not differ

TABLE 15 Employment Characteristics Among Persons with and Without Disabilities, with and Without Adjustments for Age and Gender, California, 1999

Employment Characteristic	All Persons	Unadjusted		Age and Gender Adjusted	
		Without Disability	With Disability	Without Disability	With Disability
All-employed, age 18–64, n	1,220	1,099	121	—	—
Self-employed (percent)	13.6	13.2	17.4	15.1	18.3
Working day shift (percent)	77.5	77.9	73.6	79.8	73.6
Flexible hours (percent)	55.2	55.6	52.1	55.5	52.8
Work at home all the time (percent)	5.9	5.4	9.9[a]	6.3	10.2
Contingent employment (percent)[b]	10.7	10.5	13.2	9.8	12.7
Not permanent job (percent)	8.8	8.6	9.9	8.3	10.0
Temp agency employed (percent)	2.9	2.6	5.0	2.5	4.6
Job tenure (percent with years on job):					
One year or less	24.2	23.5	30.6	—	—
>1 to 5 years	34.8	34.8	33.9	—	—
6 to 10 years	17.9	18.2	14.9	—	—
More than 10 years	22.9	23.1	20.7	—	—
Less than 5 years on job (percent)	53.1	52.4	59.5	45.9	56.8[a]
Less than 1 year on job (percent)	19.3	18.7	24.8	15.7	24.2[a]
Job tenure, mean	—	6.8	6.2	8.0	6.5[a]
Psychological characteristics of jobs					
Required to learn new things	89.5	88.9	94.2	89.1	94.6
Has little freedom to decide how to do work	25.2	25.2	24.8	23.9	24.9
Makes a lot of decisions on one's own	82.0	82.0	82.6	83.4	83.0
Has enough time to get job done	77.5	77.8	75.2	76.8	74.6
Required to work very fast without breaks	43.0	43.4	38.8	42.6	38.9
High-demand, low-control job[c]	14.7	14.9	11.6	11.7	14.6

[a] $p < .05$.
[b] Contingent employment includes nonpermanent workers and temporary agency employees.
[c] A job is considered to be high demand and low control if the respondent states that he or she has little freedom to decide how to do the job, and either does not have time to get the job done or is required to work very fast without breaks.
SOURCE: Authors' analyses of the California Work and Health Survey.

significantly in the percentage reporting being self-employed, working a day shift, having flexible hours of employment, and having contingent employment, or in the psychological characteristics of jobs.

Job Displacement and Accession

The biannual Bureau of Labor Statistics survey used to establish the rate of job displacement does not include a measure of disability status. Accordingly, we use the California Work and Health Survey to analyze differences between persons with and without disabilities in rates of job loss (Table 16). In contrast to the findings with respect to working conditions, persons with disabilities reported much higher rates of job displacement than those without; adjustment for age and gender did not alter this finding. Thus, persons with disabilities were almost twice as likely to report job loss in the year prior to the survey as those without (17.0 versus 9.6 percent). Such persons were more than 70 percent more likely to report job loss in the three years prior to the survey (33.0 versus 19.1 percent). Using the federal government's strict definition of job displacement—job loss in the past three years among persons 20 and over who had held the job for three or more years—persons with disabilities were more than 75 percent more likely to have met this criterion than those without disabili-

TABLE 16 Involuntary Job Loss Among Persons with and Without Disabilities, with and Without Adjustments for Age and Gender, California, 1999

Involuntary Job Loss	All Persons	Unadjusted		Age and Gender Adjusted	
		Without Disability	With Disability	Without Disability	With Disability
All persons, age 18–64, employed within past 3 years	1,503	1,316	188	—	—
Job loss in past year	10.5	9.6	17.0[a]	8.6	17.2[a]
Job loss in past 3 years	20.8	19.1	33.0[a]	17.6	33.0[a]
Displaced[b] in past 3 years	7.0	6.4	11.4[a]	6.8	11.5[a]

[a]$p < .05$
[b]Definition of displacement used by the federal government: person aged 20 or over, with at least 3 years' tenure on job.
SOURCE: Authors' analyses of the California Work and Health Survey.

ties (11.4 versus 6.4 percent). Using the longitudinal component of the California Work and Health Survey, we estimated the proportion of persons with and without disabilities who were not working in one year who had become employed by the time we reinterviewed them a year later. Persons with disabilities were about 61 percent as likely to enter employment as persons without disabilities (job entrance rates were 20.3 and 37.9 percent, respectively) (data on job entrance not in tables).

SUMMARY OF LABOR MARKET DYNAMICS

This review of overall trends in the labor market and of trends affecting persons with disabilities has yielded a partial description of how things are, not how they might be in the years to come. However, the major trends in employment—the decline in labor force participation among older men, the increase among younger women, the shift from manufacturing to service industries and occupations, and the emergence of new terms of employment—have been unfolding for several decades, and with the possible exception of the decline in labor force participation among older men and the end of the increase in labor force participation among women, there are no major disjunctures forecast for the remainder of these trends in the years to come (Bowman, 1997).

More importantly, this review is a description of whether persons with disabilities *do* work and, if so, *how and where*, not of whether they *can* work. However, the evidence presented in this paper is consistent with the notion that given the appropriate economic climate, a substantial number of persons with disabilities will enter the labor market and then maintain employment. Because a relatively small proportion of persons with disabilities do work and the exact proportion shifts with changes in the state of the labor market, there would appear to be a reasonable number who *could* work in the appropriate circumstances.

What is preventing them from doing so? Yelin and Trupin (2000) recently completed an analysis of the factors affecting transitions into and out of employment among persons with and without disabilities. For persons with disabilities, demographic characteristics were the principal factors affecting the probability of entering employment, with those 18 to 24 years of age six times more likely to do so than those 55 to 64 years of age and with whites 40 percent more likely to enter jobs than nonwhites. Other social and demographic factors related to job entrance among persons with disabilities included marital status, household type, education, residential environment, and baseline household income; gender, Hispanic ethnicity, and region of the country were not associated with job entrance in this group. Demographic and social factors associated with maintaining employment included age, race, gender, marital status, education, region, and baseline household income; Hispanic ethnicity, residential environment, and household type were not associated with maintaining employment among persons with disabilities. Interestingly, the principal *work-related* factor affecting whether persons with disabilities maintained employment was the industry in which they worked, whereas the principal work-related factor affecting whether persons without disabilities did so was their occupation. This suggests that the probability that persons with disabilities will be able to keep working after the onset of impairment is determined to a large extent by the welfare of the industries in which they work, rather than their own characteristics. The welfare of persons without disabilities, in contrast, is tied to a greater extent to their personal background. Expanding industries will find a way to accommodate the needs of their workers with disabilities, level of impairment notwithstanding.

Thus, the question of how to assess functional capacity for work cannot be asked abstractly. Instead, it must be asked assuming a strong demand for labor and the presence of reasonable accommodation, as mandated by the Americans with Disabilities Act of 1990 (West, 1991). Nevertheless, even when these conditions are met, many individuals will not be working, suggesting that it may be possible to describe a core set of functional requirements that apply even when the demand for labor is strong. Although the capacity to "tote that barge and lift that bale" still applies to some jobs, increasingly the core competencies would appear to revolve around the ability to communicate, concentrate, interact with others, learn new tasks, and be flexible in how and with whom work gets done (Osterman, 1988). This is true even when a job demands the capacity for toting and lifting, but it is especially true in the growth sectors of the economy in which the physical demands of work may be minimal.

MEASURING FUNCTIONAL DEMANDS OF THE CONTEMPORARY AND FUTURE LABOR MARKETS

O*Net[6] (Occupational Information Network) has been developed under a contract from the Department of Labor to replace the Dictionary of Occupational Titles (DOT) as the principal way of assessing the functional demands of jobs (Peterson et al., 1996). The purpose of O*Net was twofold: (1) to create an on-line database of work requirements in order to provide job information in an accessible format and one that can be readily updated, and (2) to provide a listing of job characteristics that reflect the

---

[6]This discussion is based in part on a discussion with our colleague Ms. Katie Maslow, but any errors of fact or interpretation are our own.

contemporary economy. The DOT characterized jobs on the basis of the complexity of dealing with data, people, and things. O*Net characterizes both the attributes of occupations and the characteristics of the people who fill each job. Data are collected on six separate dimensions: (1) experience requirements (training, experience, licensing); (2) worker requirements (functional skills, general knowledge, and education); (3) worker characteristics (abilities, interests, and work styles); (4) occupation characteristics (labor market information, occupational outlook, and wages); (5) occupational requirements (work activities, work context, and organizational context); and (6) occupation-specific information (knowledge required to do an occupation, occupational skills, and the specific tasks on the job). The data for O*Net derive from a survey of job analysts and from interviews with persons in each occupation (The latter source will include a greater number of characteristics than the former, but the data will be available later.). In both instances, respondents will be asked to report the level of each characteristic on a scale; the average level among all respondents for each characteristic will be disseminated.

A thorough description of O*Net and of how it may be used is beyond the scope of this paper, as is a listing of its shortcomings with respect to assessment of the functional capacity of applicants for disability benefits. For the former, suffice it to state that O*Net has the capacity to capture the complexity of each job through the diversity of the dimensions measured and the heightened pace of change in the nature of each job. For the latter, suffice it to state that O*Net's principal limitation is its reliance on the average level among respondents for each job characteristic, while those adjudicating applications for disability benefits need to assess minimal requirements on each such characteristic. However, in capturing the complexity of the modern job, O*Net solves one problem for those assessing capacity for work (providing a contemporary model of work), while raising another (providing no easy method to assess which among six dimensions and 300 specific characteristics are the essential functions of a job and which, therefore, are central to an assessment of functional capacity).

Indeed, this conundrum is not unique to the situation facing those who would adjudicate applications for disability benefits. In assessing whether employers are in compliance with the employment requirements of the ADA, the Equal Employment Opportunity Commission is asked to assess whether an individual can perform a job's essential function, but the law provides little guidance in how to determine what such a function is (Jones, 1991). If we are right that an increasing proportion of jobs involve complexity and dynamism in tasks, competencies, and relationships with colleagues, then it necessarily follows that a system to assess functional capacity must take this complexity into account today and incorporate the ability to measure—if not predict—changes in these characteristics in the years to come. The jobs that can be reduced to one unvarying essential function may be those few of us want and, paradoxically, those that—because of their high levels of physical demand—few persons with disabilities can perform.

SUMMARY AND CONCLUSIONS

Retrospective assessment of past attempts to predict the future of the labor market suggests that one should be humble in trying to project the shape of employment in the years to come. Many, if not most, of the predictions of the late 1950s and 1960s proved unfounded. At that time, many analysts saw automation as the principal threat to the labor market, with rising unemployment and de-skilling of jobs the necessary result of this trend.

Today, we are concerned about the erosion of job security and we wonder how many of us can cope with the demands of the service economy (and even the manufacturing sector) for a flexible response to a varying set of tasks. However, recent projections concerning the nature of the labor market call some of our predictions about even the near future into question (Bowman, 1997). In the last several decades, the labor force has grown with the entrance of women into employment and the service sector has expanded. Attenuation of the former trend necessarily will occur: most of the women who could enter work have already done so. While the latter trend is expected to continue overall, some parts of the manufacturing sector are projected to expand, particularly industries related to exports and the manufacture of items requiring high levels of capital investment. Nevertheless, all projections for the future suggest that the premium paid to those with high levels of education will continue and that flexibility on the part of the worker will be of paramount importance.

The fears of 40 years ago proved unfounded because the only model we had to work with was a mechanistic model of the production of goods. In that model, we believed it would be relatively easy to assess capacity for work. Most of those who would apply for disability benefits were blue-collar workers in the manufacturing sectors with degenerative, largely physical conditions of aging. The fears of today may be unfounded because the majority of tomorrow's workers may function much better than our own generation in jobs with a complex and varying set of tasks and because we may learn to accommodate the needs of the minority of workers—those with cognitive and behavioral impairments—who cannot do well in this situation today.

Just as the past generation was unable to predict what the world of work would be like in the year 2000, we cannot know with certainty what jobs will demand of us in the future. However, we have learned something: that any system put into place to assess the capacity for work must accommodate rapidly changing conditions. The visionary and all-encompassing criteria of today necessarily become the mechanistic ones of tomorrow unless we build in the capacity to change the criteria as quickly as the economy evolves, which in turn requires us to have in place a strong research infrastructure to understand the changes and to develop the tools to measure them.

REFERENCES

Bell D. 1973. *The Coming of Post-Industrial Society: A Venture in Social Forecasting.* New York: Basic Books.
Belous R. 1989. *The Contingent Economy: The Growth of the Temporary, Part-Time, and Subcontracted Workforce.* Washington, DC: National Planning Association.
Benson V, Marano M. 1998. Current estimates from the National Health Interview Survey, 1995. National Center for Health Statistics. *Vital Health Stat* 10(199). [Online]. Available: http://www.cdc.gov/nchs/data/series/sr 10/10 199 1.pdf.
Berkowitz E. 1987. *Disabled Policy: America's Programs for the Handicapped.* New York: Cambridge University Press.
Bound J, Waidmann T. 2000. *Accounting for the Recent Declines in Employment Rates Among the Working-Aged Disabled.* Cambridge, Massachusetts: National Bureau of Economic Research Working Paper 7975.
Bowman C. 1997. BLS projections to 2006: A summary. *Monthly Labor Review* 120(11):3–5.
Braverman H. 1974. *Labor and Monopoly Capital: the Degradation of Work in the Twentieth Century.* New York: Monthly Review Press.
Burkhauser R, Daly M, Houtenville A. 2000. *How Working Age People with Disabilities Fared Over the 1990s Business Cycle.* Cornell University: Rehabilitation Research and Training Center for Economic Research on Employment Policy for Persons with Disabilities.
Chirikos T. 1993. The Composition of Disability Beneficiary Populations: Trends and Policy Implications. Washington, DC: U.S. Department of Health and Human Services, Office of the Assistant Secretary for Planning and Evaluation.
Clinton A. 1997. Flexible labor: Restructuring the American work force. *Monthly Labor Review* 121(8):3–17.
Cornfield D. 1987. *Workers, Managers, and Technological Change.* New York: Plenum Press.
Derthick M. 1990. *Agency Under Stress: The Social Security Administration in American Government.* Washington, DC: Brookings Institution.
Hale T. 2001. *The Federal Effort to Identify People with Disabilities in the Current Population Survey.* Unpublished paper, Department of Labor, Bureau of Labor Statistics.
Hirschhorn L. 1988. *The Workplace Within: Psychodynamics of Organizational Life.* Cambridge, MA: MIT Press.
Hirschhorn L. 1991. Stresses and patterns of adjustment in the postindustrial factory. In: *Work, Health, and Productivity* (Green G, Baker F, eds.). New York: Oxford University Press.
Hipple S. 1997. Worker displacement in an expanding economy. *Monthly Labor Review* 120(12):26–39.
Hipple S. 1999. Worker displacement in the mid-1990s. *Monthly Labor Review* 122(7):15–32.
Jacobs E, Zhang H. 1998. *Handbook of U.S. Labor Statistics: Employment, Earnings, Prices, Productivity, and Other Labor Data.* (2nd edition). Lanham, MD: Bernan Press.

Jones N. 1991. Essential requirements of the Act: A short history and overview. In: *The Americans with Disabilities Act: From Policy to Practice* (West J, ed.). New York: Milbank Fund.

Kelley M. 1990. New Process Technology, Job Design, and Work Organization: A Contingency Model. *American Sociological Review* 55:191–208.

Levine L. 2000. *The Employment of People with Disabilities in the 1990s.* Congressional Research Service. August 15. Order Code RL 30653.

Levy F. 1998. *The New Dollars and Dreams: American Incomes and Economic Change.* New York: Russell Sage Foundation.

Mashaw J, Reno V. 1996. Overview. In: *Disability, Work, and Cash Benefits* (Mashaw J, Reno V, Burkhauser R, Berkowitz M, eds.). Kalamazoo, MI: WE Upjohn Institute for Employment Research.

McNeil J. 2001. Americans with disabilities, 1997. *Current Population Reports Series* P70–73. February.

Nardone T, Veum J, Yates J. 1997. Measuring job security. *Monthly Labor Review* 120(6):26–33.

Osterman P. 1988. *Employment Futures: Reorganization, Dislocation, and Public Policy.* New York: Oxford University Press.

Peterson N, Mumford M, Borman W, et al. 1996. *O*Net Final Technical Report.* Washington, DC: American Institutes for Research.

Piore M, Sabel C. 1984. *The Second Industrial Divide: Possibilities for Prosperity.* New York: Basic Books.

Polivka A. 1996. Contingent and alternative work arrangements. *Monthly Labor Review* 119(10):3–9.

Quinn R, Staines G. 1979. *The 1977 Quality of Employment Survey: Descriptive Statistics with Comparison Data from the 1960–70 and the 1972–73 Surveys.* Ann Arbor, MI: Survey Research Center.

Schwartz J, Pieper C, Karasek R. 1988. A procedure for linking psychosocial job characteristics data to health surveys. *American Journal of Public Health* 78(8):904–909.

Stapleton D, Barnow B, Coleman K, et al. 1994. *Labor Market Conditions, Socioeconomic Factors, and the Growth of Applications and Awards for SSDI and SSI Disability Benefits.* Report prepared for the Department of Health and Human Services, Office of the Assistant Secretary for Planning and Evaluation, under Contract 100–0012.

Starr P. 1982. *The Social Transformation of American Medicine.* New York: Basic Books.

Stone D. 1984. *The Disabled State.* Philadelphia: Temple University Press.

Trupin L, Sebesta D, Yelin E, LaPlante M. 1997. Trends in labor force participation among persons with disabilities, 1983–1994. *Disability Statistics Report* (10). Washington, DC: U.S. Department of Education, National Institute on Disability and Rehabilitation Research.

U.S. Bureau of the Census. 1981. *Statistical Abstract of the U.S., 1981.*

U.S. Bureau of the Census. 1984. *Statistical Abstract of the U.S., 1984.*

U.S. Bureau of the Census. 1990. *Statistical Abstract of the U.S., 1990.*

U.S. Bureau of the Census. 1997. *Statistical Abstract of the U.S., 1997.*

U.S. Bureau of the Census. 2000. *Statistical Abstract of the U.S., 2000.*

U.S. Department of Labor, Bureau of Labor Statistics. 1985. *Handbook of Labor Statistics.* Washington DC: U.S. Department of Labor.

U.S. Department of Labor, Bureau of Labor Statistics. 1988. *Labor Force Statistics Derived from the Current Population Survey, 1948–1987.* Washington DC: U.S. Department of Labor.

U.S. Department of Labor, Bureau of Labor Statistics. 1997. *Employee Tenure in the Mid-1990s.*

U.S. Department of Labor, Bureau of Labor Statistics. 1998. *Work at Home in 1997.*

U.S. Department of Labor, Bureau of Labor Statistics. 1999a. *Annual Average Data from the Current Population Survey*. [Online]. Available: ftp://ftp.bls.gov/pub/special.requests/lf/aa99/aat3.txt.

U.S. Department of Labor, Bureau of Labor Statistics. 1999b. *Contingent and Alternative Employment Arrangements*, February 1999. [Online]. Available: ftp://146.142.4.23/pub/news.release/conemp.txt.

U.S. Department of Labor, Bureau of Labor Statistics. 2000. *Employee Tenure in 2000*.

U.S. Department of Labor, Bureau of Labor Statistics. 2001. *Labor Force Statistics from the Current Population Survey, Historical Data for the "A" tables of the Employment Situation News Release*. [Online]. Available: http://stats.bls.gov/webapps/legacy/cpsatab4.htm.

West J. 1991. The social and policy context of the Americans with Disabilities Act. In: *The Americans with Disabilities Act: From Policy to Practice* (West J, ed.). New York: Milbank Fund.

Wilson W. 1997. *When Work Disappears: The World of the New Urban Poor*. New York: Knopf.

Wright E, Singleman J. 1982. Proletarianization in the American class structure, 1960–1980. *American Journal of Sociology* 93(1): 1–29.

Yelin E. 1992. *Disability and the Displaced Worker*. New Brunswick, NJ: Rutgers University Press.

Yelin E, Katz P. 1994. Labor force trends of persons with and without disabilities. *Monthly Labor Review* 117(10):36–42.

Yelin E, Trupin L. 2000. Successful labor market transitions for persons with disabilities: Factors affecting the probability of entering and maintaining employment. In: *Expanding the Scope of Social Science Research on Disability* (Altman B, Barnartt S, eds). Stamford, CT: JAI Press.

Yelin E, Nevitt M, Epstein W. 1980. Toward an epidemiology of work disability. *Milbank Memorial Fund Quarterly: Health and Society* 58(3):386–415.

Index